Continuity in Sustainable Facades

Kyoung Hee Kim

Continuity in Sustainable Facades

Design and Engineering of Tall Building Enclosures in Hot Climates

Springer

Kyoung Hee Kim
University of North Carolina Charlotte
Charlotte, NC, USA

ISBN 978-3-032-19840-2 ISBN 978-3-032-19841-9 (eBook)
https://doi.org/10.1007/978-3-032-19841-9

This Springer imprint is published by the registered company Springer Nature Switzerland AG
The registered company address is: Gewerbestrasse 11, 6330 Cham, Switzerland

Acknowledgement This book began in 2014, during a site visit to the Shenzhen Energy Mansion in Shenzhen, China, where I started documenting the real-time construction process of a tall building enclosure. What began as on-site observations to construction methods, curtainwall types, and technical details gradually evolved into a broader exploration of tall building enclosures in hot and humid climates. Over time, this research inquiry expanded across diverse hot climatic and urban contexts to include the documentation and study of additional projects, including MahaNakhon in Bangkok and Le Nouvel KLCC in Kuala Lumpur. These site visits and field observation guided the scholarly direction of this book and also led to the expansion of additional case study analysis in Chap. 2.

Along this journey, this book was made possible through immense support, collaboration, and shared research inquiry with colleagues and research assistants. I am deeply grateful to Ibha Shrestha for her exceptional contribution to Chap. 4. Her thoughtful and professional reproduction of drawings and willingness to test and refine façade details brought clarity and precision to the technical narrative. My sincere thanks also go to Sarah Spalding and Kyah Tucker for their dedicated support in the production drawing for Chap. 2, and to Jamar Moore and Arundhati Chitnis, whose thoughtful assistance with the literature review strengthened the scholarly relevance of the analysis.

I extend special thanks to Front Inc., Sanxin Façade Technology, and Uniwall for enabling my site visits and documentation of built projects. Their generosity and professional trust made on-site observation and rigorous evaluation of built façade systems analysis.

I would also like to acknowledge the institutional support from Chris Jarrett, chair at the time this research began; Ken Lambla, then dean; Blaine Brownell, current director; and José Gamez, current dean. Their continued support of integrating research into my studio and class settings cultivated a fertile environment where research, teaching, and practice could intersect in a synergistic way.

Finally, my deepest gratitude is reserved for my family—my husband, Yau Shun Hui, and my three children, Anthony, Henry, and Patrick. Their patience, encouragement, and unwavering support sustained this work over many years and made this book possible.

Competing Interests The author has no competing interests to declare that are relevant to the content of this manuscript.

Contents

Chapter 1
Tall Buildings and Sustainability

Abstract High-rise towers present both advantages and disadvantages when compared to low-rise development and horizontal urban construction. Higher population density and compact building forms in cities offer advantages such as the conservation of natural and green areas within the urban settings and the reduction of the extensive spread of infrastructure and associated social amenities. A lower ratio of building envelope area to floor area or small enclosure perimeter can help reduce construction costs, material use, and heat loss or heat gains through the building envelope. On the other hand, tall buildings may limit users' experience of the natural environment and, in some cases, create a sense of disconnection from street-level activity. Some new tall buildings and redevelopment projects are designed as iconic landmarks and prioritize aesthetic expression over sustainable performance, cultural responses, local identity, or the psychological needs of inhabitants. This section discusses how these challenges are not inherent to vertical urbanization but can be improved through sustainable design and operational decisions as well as policy and regulatory enforcement that promote culture, history, and locality. Incorporating circular strategies such as design for deconstruction, material recoverability, as well as adaptive reuse and retrofitting can significantly lower environmental impacts across the building life cycle. These strategies are most effective when applied through systems integration between building enclosures, environmental control systems, and structure systems, allowing them to function synergistically as an integrated whole rather than an isolated building system. Overall, the integration of nature and on-site renewable energy systems, the promotion of social interaction, circular economy practice, and enhancement of resilience in tall buildings can offset the environmental, economic, and social challenges.

Keywords Sustainable urbanization and tall building enclosures · Circular economy · Building life-cycle performance · Social sustainability · Urban resilience

K. H. Kim, *Continuity in Sustainable Facades*,
https://doi.org/10.1007/978-3-032-19841-9_1

1.1 Tall Buildings and Circularity

1.1.1 Urban Densification and the Rise of Tall Buildings

The growing influx of urban populations results in the densification of tall buildings in cities. Contemporary tall buildings have adopted a wide range of glazed skins due to their lightweight and contemporary aesthetics. Beyond the formal expression, tall buildings reflect various attributes such as cultural, environmental, and economic aspects inherent to the building and its context. In other words, tall towers are a synthesis of various design criteria and functional requirements, including local culture, site context, architectural programs, environmental stewardship, construction techniques, budget constraints, building code compliance, and so on, leading to strong formal implications.

The Council on Tall Buildings and Urban Habitat uses 200–300 m in height for their tall building studies, 300–600 m in height for supertall, and over 600 m in height for megatall building studies [1]. The International Building Code (IBC) defines a tall building as any building with an occupied floor more than 23 m (75 ft) above the lowest level of fire department access [2], whereas super high-rise buildings are defined by those taller than 128 m (420 ft) in height, where more onerous life safety measures are required [3]. American Society of Heating, Refrigerating, and Air Conditioning Engineers (ASHRAE) defines a tall building as taller than 91 m (300 ft), super tall as taller than 300 m (984 ft), and megatall as taller than 600 m (1968 ft). Table 1.1 shows the different height criteria that define tall towers. Varying heights represent broader values such as life safety, environmental control strategies, urban identity, economic constraints, and so on, and tall buildings reflect inherently multifaceted considerations.

By 2030, two out of every three people worldwide will live in cities [10], leading to an increase in tall building construction as a common building typology in city centers. Tall buildings offer an opportunity to integrate the concept of "urban mining", in which circular economy principles are applied to urban buildings toward more waste-free, resource-efficient, and environmentally responsible cities. This approach adopts closed-loop material flows where valuable resources can be reclaimed and reused, promoting circularity practices throughout a building's life cycle [11].

1.1.2 Circular Economy Principles in Tall Building Design

Design for Deconstruction and Reuse: Both tall and low-rise buildings can incorporate building systems and components that allow for effective disassembly and reuse after demolition. The modularity and connection design are key design criteria to achieve a circular economy. For example, prefabricated modular wall systems with reversible connections enable not only fast installation and reduced waste generation

Table 1.1 Various tall tower definition

Standards/codes	Designation	Height	Notes
ASHRAE Technical Committee for Tall Buildings [4]	High-rise Supertall Megatall	> 91 m (300 ft) > 300 m (984 ft) > 600 m (1968 ft)	Vertical transportation and environmental control strategies
International Building Code [5]	High-rise	> 23 m (75 ft)	Accessible length by ladders used by fire fighters
CTBUH [6]	Tall Supertall Megatall	200–300 m (656–984 ft) > 300 m (984 ft) > 600 m (1968 ft)	City branding and architectural innovation
Emporium [7]	High-rise	35–100 m (115–328 ft) high or with 12 stories at minimum	Global skyline comparison
Leicester City Council [8]	High-rise	> 20 m (66 ft)	Accessible length by ladders used by fire fighters; city zoning height tied to historic context
German regulations [9]	High-rise	> 22 m (72 ft)	Accessible length by ladders used by fire fighters

but also easy upgrades of components and future reuse. The connection design that links different building materials or systems is another important design factor toward circular economy. Manufacturing standardized systems with easily separable joinery reduces the effort needed for installing, disassembling, and reassembling building systems or components. In other words, using mechanical joints instead of adhesives makes architectural components easier to separate for transportation, installation, and disassembly, thereby supporting circular resources and extending the longevity of useful and valuable architectural materials. Promoting the disassembly and reuse of building components, as well as increased longevity of building materials, is important for achieving a circular economy in building construction. For example, tessellated structural-architectural (TeSA) walls enhance the reuse and structural integrity of building systems by using interlocking tiles designed as repeating, interchangeable modular units, leading to simple prefabrication, dismantling, and reconstruction [12]. "Reversible connections" improve building material circularity by enabling easy assembly, disassembly, and reuse of systems and components, which can be reused within the same building or in other buildings [13]. The concept of reversible connections can apply to other building systems, such as enclosures, services, and non-structural components, as well as building modules connected with reversible connectors.

Material Recovery and Recycling: The use of recyclable and reclaimed materials for high-rise building construction mitigates adverse effects on the environment. Recent studies on embodied carbon of tall buildings highlight the importance of material selection and reuse. For example, substructure, primary structures, and building enclosures of high-rise buildings account for 85 ~ 95% of the total embodied carbon, whereas internal partitions and interior systems take a relatively smaller portion [14]. The recycling of concrete and the recycling/reuse of steel decrease raw production and processing. Reclaiming heavy timber and/or cross-laminated timber (CLT) lessens the exploitation of new wood. This recycling, reuse, and reclamation practice enables a reduction in the carbon dioxide footprint associated with the construction and operational activities of tall buildings and limits the exploitation of materials. A comparative study indicated that the life cycle CO_2 emissions from a CLT building show a negative impact of − 210 kg CO_2/m^2 at year 1 and − 174 kg CO_2/m^2 at year 100 due to its carbon storage capability, whereas a reinforced concrete building shows 170 kg CO_2/m^2 at year 1 and 221 kg CO_2/m^2 at year 100 due to operation and maintenance [15]. This circular material approach improves the life cycle sustainability of tall buildings, which in turn supports their long-term value. Additionally, setting up and defining material passports and inventories used in existing buildings can enhance material recovery and recycling [16]. Material passports contain information on the place of origin, the type of materials and their quantities, and their texture and quality, which are necessary and sufficient for traceability and further recycling of building materials. This practice can provide guidance and trust during the dismantling of elements for reuse and other uses, such as reconstruction in future projects. They engage stakeholders, designers, developers, manufacturers, and recyclers and implement sharing of resources and a circular economy. Initiatives for cooperation on material passports and sharing platforms supporting the recycling of waste materials and material exchanges promote a circular economy in the construction sector.

Product-as-a-Service (PaaS) Models for Tall Buildings: Tall buildings have more opportunities for innovation and effectiveness because cities aim to meet sustainable development and manage increasing urban populations and economic activities. Product-as-a-Service (PaaS) Models replace the focus from ownership of technology to service-based contract systems. PaaS approach can promote the technological novelty and sustainability of tall buildings throughout their lifecycle by extending performance contracts to the entire life cycle including the end-of-life period. This lifecycle service has been shown to enhance efficiency in resource consumption and waste minimization in other fields through service provider's full management centered on service- or performance-based contracts. For example, developing procedures that allow for Facade-as-a-service contracts for tall buildings can help incentivize the implementation of a circular economy during design and engineering of building enclosures and their components [17]. Facade-as-a-service includes the operation, maintenance, upgrade, and eventually recycling of façade systems, leading to more resource utilization and less wastage. The deep innovation of building envelope improvement also incorporates the integration with an electricity-based HVAC system for improved energy efficiency in real-time. Service companies offer on-site and/or remote diagnosis and predictive maintenance for facades and/or HVAC systems. When it comes to an end-of-life cycle, the service provider will prioritize the extension of service life and material recovery before proper recycling and/or disposal of the system. The service provider will collect the systems that have become old, outdated, or have been put out of service by the clients and ship them for recycling to specially developed complexes. This PaaS approach not only reduces the amount of waste generated by building systems but also enables to retrieve and recover useful materials from the systems for further use, facilitating a circular economy.

Adaptive Reuse and Retrofitting: There are many reasons for obsolescence such as the physical performance of tall buildings, changes in social needs and priorities, improvements in technology, economic shifts, and building operational features [18]. Re-purposing and retrofitting strategies make it possible to enhance the service life of existing structures' and postpone the construction of new ones, an important aspect of the circular economy. There are many low-performing buildings that lack modern technologies such as IT and HVAC, which make them less competitive in the market. Physical damage as well as poor maintenance lead to depreciation and require physical and technological updates. Alternative uses can yield profit and meet the clients' requirements if they offer healthy and comfortable environments. As a way of reducing obsolete buildings, transforming older rentable office towers into modern adaptive reuse projects can provide healthy environments, lower environmental impacts, and support economic prosperity [19]. Tall buildings offer great opportunities for promoting vertical urbanization as well as incorporating a circular economy approach. Emerging paradigms such as designing for circular economy, adaptive reuse, material recovery, and Product-as-a-Service approaches will assist in advancing the sustainability of tall buildings.

1.2 Sustainable Tall Buildings

The $I = PAT$ framework explains the interdependent and interwoven relationship between population growth (P), goods and service consumption per capita (A), and technological advancement (T) affecting environmental pollution and resource depletion. P, A, and T are interdependent and influence each other. The growing population has excelled the efficiency and manufacturing of technology to produce more goods and services, thereby increasing environmental stress and per capita consumption. The PAT framework can also explain the environmental impacts from the building sector. We create our own indoor environment by heating, cool, and ventilating the space with man-made machines independent of outdoor environments. More urban population requires more buildings to live, work, and play. Improved lifestyle leads to the construction of larger spaces furnished with bigger HVAC machines and daily commodities. Tall buildings in the U.S. take only a fraction of the total commercial building stock but consume a substantial portion of the total energy use [20]. Power production to support tall building operation consumes enormous fossil fuels and causes environmental destruction. High-density developments accommodate thousands of people in a tall building with a very small footprint. Cities like Dhaka and Mumbai have a high density of approximately 44,000 and 32,000 people/km^2 respectively, whereas New York and Paris are about 1800 and 3900 people/km^2 [21].

In a period of globalization, it is estimated that more people will live in cities than rural areas. According to the UN report, "urbanization is a complex socio-economic process that transforms the built environment, converting formerly rural into urban settlements, while also shifting the spatial distribution of a population from rural to urban areas" [22]. Tall buildings that started in Chicago and New York spread across the U.S., defining the urban cores of cities with a cluster of tall buildings surrounded by low density buildings. While spreading across the globe to parts of Asia, the Middle East, and Africa, tall buildings in the current period are proposed and constructed predominantly for economic returns rather than social welfare reasons. The number of tall buildings has increased dramatically since the 2000s, with over 1300 buildings worldwide exceeding 200 m in height in 2017 [23].

Growth in urban populations has resulted from both the increased world population and increased shift of people into urban areas. As urbanization increases, environmental stewardship is an urgent need for high rise buildings due to their significant carbon footprint. Energy codes push tall buildings to be more climatically responsive, shaping their forms and aggressively regulating the use of energy to operate them. Green roof and roof top solar requirements incorporated into the New York City codes mandate the installation of such sustainable technologies [24]. Tall building surfaces could offer much more with their height to augment sunlight for on-site energy production. Zoning codes mandate that a tall building offer daylight access on the street level. In 1916, the 'Zoning Law' in New York City regulated the setbacks of the building form toward the top of high-rise buildings to provide daylight access to pedestrians on the street [25]. This reinforces environmental design values by promoting both resource efficiency and the quality of users' daily experience. These regulations and policies shape building forms and material selection. Buildings reflecting different needs in our current time can contribute to transforming cities in more multivalent and positive ways.

Tall towers also need to make connections to the ground at their street level and to its surroundings, inviting pedestrian access and community engagement, allowing users to inhabit the building surface and the city at the same time. A design approach that develops a specific response to each project condition results in unique form-making for any type of high-rise buildings. For example, occupiable stepped building surfaces on the exterior, hanging gardens, or rooftop gardens provide opportunity to make connections to public and replicate street activities. Vertical surfaces can enrich urban experience and social interaction by functioning as extensions of the public realm, especially when they integrate multilayered interior public spaces within meso-scale building clusters [26].

1.2.1 Economic Sustainability of Tall Buildings

Major cities in the twentieth century promoted taller and more compact development because urban spaces were limited and even the interstitial spaces between the towers started to be filled in. The effects of such rapid urban population increases largely overwhelmed the capacity of infrastructure and became an additional cause of public housing shortages. Urban dualism can also be described in terms of the spatial expression of wealth—including the two types of urban lifestyles that oppose each other—the tall apartment and social housing. Tall apartments are sold first to higher income groups where public housing is readily available.

Generally, new skyscrapers are built together with additional public use activities like sky decks, community gardens, and eateries located on rooftop terraces, in which such public amenities encourage social activities. Many skyscrapers also offer viewing platforms at the upper levels. The sky deck with glass panels was first implemented at the CN Tower in Toronto in 1994, which then a series of skywalk implementations for tourism developed rapidly [27]. Very high-rise structures enhance tourism and demonstrate economic impacts on cities, following a pattern similar to the Bilbao-effect.

Construction activities can create many jobs at varying skill as well as technical and professional sector levels. Tall buildings can support social and economic sustainability by clustering businesses and creating employment [28]. The clustering of businesses can foster additional economic prosperity because other firms and developers will invest in infrastructure and other business activities, thereby promoting new business ventures as well as regional competitiveness. Previous research indicates that super tall buildings in large cities positively enhance the innovation of firms within 1 km of skyscrapers through development and knowledge sharing between the skyscrapers and surrounding firms [29].

It is common knowledge that construction costs are usually associated with time, labor and raw materials bid, and the quality of construction. These attributes affect the viability and sustainability of the project in question. Longer periods of construction tend to increase costs due to labor, scheduling, and delays in occupancy. Prefabrication, modular construction and efficient scheduling can lead to reduced timelines and overall cost reductions [30]. Labor use is very important since skilled workers tend to be expensive. Improvements in the construction cost are closely related to the management of the workers and their skills. Good working conditions assist in the retention and hiring of skilled manpower which raises productivity and eventually reduce construction cost.

Design and engineering decisions made during the design phase have significant cost implications. Tall buildings are exposed to significant forces acting in lateral directions and need to incorporate aerodynamic principles. The layout and types of structural elements such as the core geometry, shear walls, bracing, or outrigger systems occupy usable and/or rentable spaces, and reduced rentable and selling area makes it difficult to make profits [31]. Thus the shape, dimensions, and location of the structural members of the tall building become crucial not only to structural stability itself but also to the amount of floor area that can be occupied. Therefore, the optimization of structural systems is very important design and engineering process not only in resisting lateral and gravity forces but also in minimizing the space required for structural elements. The integration of computer software allows for iterative design and structural analysis process for different structural options and ensures a high level of accuracy in evaluating spatial and structural implications upon completion. The vertical structural cores offer sufficient space for their vertical transportation means, such as elevators and fire-safety stairs, as well as risers and space for MEP (mechanical, electrical, and plumbing) systems.

Another economic metric of a tall building is the net-to-gross ratio. The net-to-gross ratio is the ratio of net floor area to gross floor area and plays a critical role in the design and development of tall towers. A higher ratio means more space for building program uses. The net floor area-to-gross floor area ratio differs based on the structure system and service core layout, such as the structural strategy chosen and the degree of service systems provided to operate the high-rise towers. The structural core usually houses service systems, including vertical transportation systems, HVAC (Heating, Ventilation, and Air Conditioning) plant and distribution systems, life safety egress such as fire stairs, fire protection systems such as standpipes and sprinklers, restrooms and plumbing systems, and electrical/IT systems, which are crucial in determining the overall efficiency and space zoning in tall towers.

Another fundamental aspect of the design is to efficiently run the HVAC plant (e.g. boiler, chiller, air handling unit) and the distribution system (e.g., duct, pipes) to ensure that the spaces are well conditioned to maintain the pleasant indoor environment and user comfort. With increasing heights of buildings, air distribution reaching the upper floors to deliver conditioned air and ventilation requires larger spaces for equipment and distribution, reducing the net-to-gross ratio. Integrating innovative, energy efficient systems such as dedicated outdoor air system (DOAS) along with refrigerant based VRF systems requires less space due to their high energy efficiency, thereby allocating more space for usable working and living spaces.

In addition to the influence of building systems on net-to-gross ratio, the building enclosure-to-floor surface ratio is another important cost efficiency metric in tall building design. The number of building stories in relation to the building envelope to floor ratio is influenced by the building shape and slenderness ratio of the tower. A higher envelope area per unit floor area results in more expensive construction costs because a higher envelope-to-floor ratio requires more enclosure materials, more labor, and stronger structure needed to build more enclosures. Although a higher envelope-to-floor area requires high upfront costs, improving the external facades may help increase the number of outdoor views available to users and daylight illumination into the space, as well as enhance the overall appearance and market value of the building. This is because people are willing to pay a premium for buildings with windows that have remarkable views [32]. This is especially significant for residential properties since most buyers prefer to have panoramic landscapes or beautiful city views through their windows. Therefore, the challenge of design is to balance the additional structure and up-front cost which entails providing better views, more daylight, quality experience, and better aesthetic characters. Some stakeholders may consider the benefits from improved daylighting and better property values to outweigh the increased upfront costs.

Additionally, there are economic benefits that can be derived from sustainability branding for high-rise buildings. The literature shows a positive relationship between green building certification and the healthy building market valuation,

Table 1.2 Economic sustainability of tall buildings

Project life cycle	Economic considerations of tall buildings
Design and engineering phase	Balance between exterior facades and benefits Maximum net to gross area considering the balance between structure, transportation, MEP system, life safety system, and space utilization
Construction phase	Construction cost—schedules and labor management Job creation Investments in infrastructure and urban development
Operation phase	Iconic visiting tourist spot
End of life phase	Reclaimed building components and adaptive reuse for circular economy

rent income, and sale prices [33]. Greener designs and technologies enhance developers' marketability, enabling them to attract selective tenants as well as retain and recruit talented employees. As sustainability branding of high-rise buildings offers an opportunity to position sustainable buildings as a green icon within the community, it allows these buildings to play more active roles in their local contexts.

These buildings can achieve net zero operational energy performance including energy efficiency, renewable energy systems, water use reduction strategies and robust waste management systems that help reduce their carbon emissions compared to conventional buildings. Developers, building owners, and tenants who make a greening efforts and use sustainability branding effectively, tend to engage broader audiences with diverse interests, raise awareness, and provide positive leadership within the socio-cultural context. Table 1.2 outlines the economic considerations for each project life cycle stage.

1.2.2 Ecological Sustainability of Tall Buildings

The different perspectives on contemporary urbanization affect urbanization's future. Supporters of tall building urbanization continue to develop higher density cites because tall towers not only prevent urban sprawl but also reduce the costs of urban infrastructure, land utilization, and urban green space. Both high urban density and the compactness of tall buildings can reduce energy use associated with heating, cooling, ventilating, and lighting, leading to lower per capita energy consumption [34]. From another perspective, high density urban development demands much higher embodied energy and environmental burdens from intensive construction activities as well as continuous operation and maintenance. This means that high-rise buildings have higher embodied carbon and operational carbon over the life cycle of the building. This negative effect in terms of the environment impact of tall buildings might outweigh all positive aspects. As a result, it is important to holistically evaluate design measures when introducing high rise structures and high density development without causing drastic changes to urban ecosystems, urban heat island effect, and alterations to the local environmental quality [35].

Many examples worldwide demonstrate that an increase in the number of tall buildings reflects positive trends in urbanization. Cities such as New York and Hong Kong are famous for their iconic skyscrapers. On the other hand, megacities such as Mexico City and Mumbai have a very few tall buildings. It is evident that the high-rise and low-rise development reflects the complexities of urban planning and development shaped by historical, socioeconomic, regulatory policies, and cultural conditions [36]. All these different aspects of sociological and cultural factors come together and determine urban verticality in different urban areas.

The life cycle of the tall building can be examined using pre-use, use, and post-use phases. The pre-use phase involves all activity associated with the resource extraction and processing, manufacturing, transportation, and construction processes. The use phase involves the functional activities of the building such as operation, maintenance, and renovation. Operational energy consumption includes heating, cooling, lighting, and ventilation systems as well as plug loads, security/IT, and vertical transportation. The post-use phase covers the demolition of buildings and end-of-life processes, which include energy consumption and carbon emissions mostly from the demolition, transportation, and landfilling activities. All these phases burn excessive energy and generate waste and pollutant emissions to air, soil, and water. Although tall buildings provide opportunities for increased efficiency, it is important to evaluate green design strategies and sustainable technologies across the holistic life cycle phases, including design, construction, operation, management and end-of-life cycle. These life cycle strategies can be achieved through high-performance building envelopes, energy efficient HVAC systems, the utilization of renewable energies and water saving measures, and the use of green materials throughout the entire life cycle of the building.

Bioclimatic design refers to designing a building to respond to the climate and the geographical conditions of the building site. It has been found that tall buildings with bioclimatic design strategies are more energy-efficient and ecologically sustainable. With regard to solar exposure in various climates, in hot climates, shading strategies such as overhangs, fins, and screens can be employed to reduce the amount of heat entering into the building, which in turn reduces the amount of cooling required to condition the building. In colder regions, tall buildings could maximize winter solar exposure and the amount of daylight penetration deep inside the building, thus reducing heating energy in winter. Bioclimatic design also incorporates facade orientation, size and arrangement of the windows, and high levels of thermal insulation. The tall building enclosures serve as a prime location for harnessing solar energy and regulating energy flows during the day. Energy efficient tall buildings with bioclimatic design strategies can achieve net-zero energy buildings by using collected solar energy to power building operations. Tall buildings also affect the amount of sunlight received by other neighboring buildings, which in turn changes the microclimate around these buildings, which may consequently impact their energy usage, daylighting, and thermal comfort levels within the buildings. The bioclimatic design coupled with on-site renewable energy generation, can reduce energy demand as a primary strategy while offsetting remaining needs through clean energy production.

High-rise structures with poorly designed envelope configurations, shapes and surface treatments can cause glare and thermal stress on neighboring structures and streets. In addition, tall buildings can also cause overshadowing and create negative impacts on the neighboring properties by reducing the amount of natural lighting received and decreasing solar availabilities. These negative impacts can have down-stream effect on building energy use, pedestrian comfort, indoors environmental quality, and the aesthetics of urban spaces, especially in high-density tall building districts. When properly designed, tall buildings can be aimed to improve the built environment in a way that moderates microclimatic parameters on streets, pavements and public spaces, thereby enhancing the microclimates while reducing the need for building energy consumption.

Tall buildings can have a significant impact on the amount of sunlight received by surrounding buildings, influencing their microclimate and potentially affecting their energy bills, daylighting, as well as thermal and visual comfort levels. Several built examples have shown that the building form, façade geometry, and surface properties of tall buildings can magnify solar intensity, causing increased temperatures and glare problems in nearby buildings and streets. On the other hand, the overshadowing effect of tall buildings can result in reduced access to natural light, increased shading, and diminished solar exposure for neighboring properties. These overshadowing or "death-ray" phenomena may have implications for building energy consumption, pedestrian comfort, indoor comfort, and the visual environment, particularly in dense urban areas where tall buildings are prevalent. Building facades can be holistically evaluated and strategically designed to minimize adverse microclimatic conditions on streets, sidewalks, and public spaces, thereby creating more comfortable outdoor and indoor environments in dense urban fabrics and reducing the demand for building energy consumption.

Heat retention in urban areas resulting from urban canyon effect and low-albedo materials causes urban heat island (UHI) effect, leading to higher temperatures in urban areas than in rural areas. This condition is worsened by building surface materials (e.g. low surface albedo, asphalt and concrete), surface configurations (e.g. urban canyons from tall buildings) and human use such as transportation, energy used by buildings (e.g. compressor, cooling tower), and on-site fuel combustion [37]. The low-albedo surfaces of buildings, roadways and other infrastructure absorb and hold heat in the environment. Air conditioners in the building discharge a huge amount of waste heat into the surrounding environment. High-rise buildings can block natural airflow and trap heat in the urban space. These heat island effects demand higher cooling load to counteract the hot outdoor temperature and keep the indoor space comfortable. There is a clear correlation between urban heat islands and increased demand for cooling energy in dense cities [38]. Apart from temperature and cooling demand increases, the UHI effect can worsen the air quality levels because traffic pollutants become trapped within urban space [39]. High air pollution levels will require highly efficient air filters for mechanical ventilation, which increase both energy consumption and operational costs. Constant exposure to polluted air or insufficient air circulation may negatively affect human health and limit productivity levels. Research shows that the raised temperatures and accumulated pollutants from the UHI effect pose the greatest risks of heat stress and poor air quality to older populations, children, and individuals with any pre-existing health condition [40]. Hence, this emphasizes the need for effective strategies to mitigate the UHI effect.

Nature-based solutions: Most metropolitan areas try to go upwards and accommodate more population into a small urban area. In this aspect, integrating greenery into urban spaces can be beneficial to both ecological and social enhancement. Since urban infrastructure has insufficient ground-level green space, vertical construction can be supplemented with natural systems or biophilic elements, such as terrace balconies, sky gardens and/or phyto-veils. The incorporation of this nature's components can reduce harmful atmospheric pollutants, urban heat island effect, stormwater problems, and urban flooding. Building owners can have profitability by saving energy bills, improved employees' productivity, the creation of healthier and productive workplaces, as well as psychological and physical wellness. The strong benefits from nature-building integration facilitate the effective incorporation of nature elements within tall towers, which can outweigh challenges such as higher upfront cost and additional maintenance requirements.

Building users and communities can benefit from urban farming programs or vertical gardens through the use of three-dimensional vertical green space. Vertical applications of vegetative cover or plants are arranged in planters or hydroponic or soil-based planter systems which are supported by the structural systems of the façade. The vertical green system needs periodic maintenance from either maintenance catwalks or building maintenance unit (BMU) to access the system. They are acceptable for use in a closed-loop systems with rainwater harvesting where overuse of portable water is limited. Sky gardens are green spaces located at the intermediate levels and rooftops of tall buildings to enhance the available green areas within a building in highly urbanized cities. They can not only mitigate UHI effect but also enhance cooling effect from trans-evaporation from green space. In addition, green roofs provide additional thermal insulation values and limit heat transfer from between indoor and outdoor environments. They enhance energy efficiency in buildings during summer by controlling the heat surrounding the city and reduce heating load during winter by limiting heat flow from indoors (hot) to outdoors (cold). The effectiveness of green roofs also depends on plant selection and roof material durability, particularly when they serve as a leisure space for social amenities and property values. The irrigation system can be designed to utilize harvested rainwater in a closed loop. Compared with sky-garden, terraced gardens are small-scale gardens built along the perimeters of several floors and extending vertically upward. They also can be integrated into raised plant benches, walls, fences, balustrades, suspended walkways, and terraces. The presence of a terraced garden helps create an interesting three-dimensional green network across the building enclosure and urban fabrics. The use of terraced gardens, like other green strategies, helps alleviate microclimate conditions around the towers and enhances the natural environment of densely populated cities. Table 1.3 discusses policies, laws, and credits related to the deployment of green roofs and vertical green systems in the United States and worldwide.

The 2007 Chicago zoning ordinance recognizes green roofs' contribution to environmental sustainability and quality of life. By offering incentives, it encourages the integration of these eco-friendly spaces that provide public benefits such as air purification, mitigation of the urban heat island effect, promotion of biodiversity, and the supply of tranquil recreation areas. The ordinance's floor area bonus incentivizes developers to implement green roofs. Buildings are eligible for floor area bonuses when the following criteria are met: (1) a green roof covers at least 50% of the net roof area or 2000 ft^2, whichever is greater; (2) documentation shows the roof can support the required load and has adequate soil depth; (3) there is sufficient space for future installations; (4) plant varieties, soil depths, and growing media comply with guidelines; and (5) vegetation is properly maintained and subject to periodic inspections [43].

San Francisco's Better Roofs Ordinance of 2017 requires that new construction dedicate at least 15% of roof space to solar installations, or alternatively, 30% to a Living Roof (vegetated roof), or a combination of both. The Living Roof option allows the replacement of required solar installations with a 2:1 ratio of Living Roof area to solar area. The Better Roofs requirements apply to new construction projects that meet the following criteria: non-residential with a gross floor area of over 2000 ft^2, or any residential buildings of any size with 10 or fewer occupied floors [44].

New York City's Local Laws 92 and 94 of 2019 require all new buildings and alterations of existing buildings where the entire existing roof assembly is replaced to have a green roof and/or solar roof covering 100% of the roof area. The Roof Law mandates either a solar photovoltaic system generating at least 4 kW, a green roof system, or a combination of the two. Vertical extensions or horizontal enlargements must also comply with these requirements [45].

The City of Toronto's 2009 Green Roof Bylaw requires green roofs for new construction and retrofit projects, with the required coverage ranging from 20 to 60% of the available roof space, depending on the building's gross floor area. The bylaw mandates green roofs for new commercial, institutional, and residential developments with a gross floor area of at least 2000 m^2, floor additions where the new gross floor area exceeds 2000 m^2, and industrial buildings with a gross floor area of over 2000 m^2 [46].

Hong Kong's BEAM Plus is a green building certification program administered by the Hong Kong Green Building Council (HKGBC) since 2010. BEAM Plus has expanded beyond Hong Kong, and its tools continue to evolve to align with international trends and standards. The program offers green certification for sustainable buildings. High BEAM Plus ratings can be achieved through integrating green roofs of nature [47]. HK Electric's Smart Power Fund, for example, incentivized energy efficiency improvement for buildings. The Water Supplies Department offers free tap flow controllers under the "Let's Save 10L Water" Campaign. In addition, financial incentives such as Green Mortgage Plan support sustainable buildings and promote a greener future [48].

Table 1.3 Policies, laws, incentives for green roofs and green walls in the US and the world [41, 42]

	City	Law/policy title	Year enacted	Operative provision
Green roof mandates	Basel, Switzerland	Building and Construction Law [4]	2002	Green roofs are mandated on all new construction and renovation projects involving flat roofs
	Copenhagen, Denmark	Green Roof Policy	2010	Mandates green roofs on all new buildings with roof slopes of less than 30°
	France (nationwide)	Biodiversity Act and Green Roof Statement	2015	Developers are required to build either green roofs or solar panels on all new commercial buildings
	Munich, Denmark	Green Roof Ordinance	1997	Mandates green roofs on all flat roofs larger than 100 m^2
	Portland, OR	Ecoroof Requirement	2018	Mandates green roofs to cover 100% of the roof area on new buildings with a net building area of 20,000 ft^2 or more; requires green roofs to cover up to 40% of the roof area on rooftop parking lots, with allowances for several equipment and green alternatives
	San Francisco, CA	Better Roofs Ordinance	2017	Green roofs, solar panels, or a combination of both must cover between 15 and 30% of the roof area on most new construction projects. San Francisco was the first city in the U.S. to require the integration of green roofs and solar panels on new buildings
	Tokyo, Japan	Green Roof Law	2001	Green roofs are mandated on new buildings with flat roofs of at least 1000 m^2
	Toronto, Canada	Green Roof Bylaw	2009	Green roofs are required to new development or addition with a gross floor area greater than 2000 m^2 for new commercial, institutional, and multi-family residential development The required size of green roof varies from 20 to 60% of the available roof area, depending on the size of the building

(continued)

Table 1.3 (continued)

	City	Law/policy title	Year enacted	Operative provision
	Zurich, Switzerland	Green Roof Program	1991	All newly renovated buildings with flat roofs, excluding roof terraces, must incorporate green roofs
Sustainability mandates	Chicago, IL	Sustainable Development Policy	2004	Development projects receiving financial assistance or special approvals must achieve a specified number of points, which are awarded based on the implementation of specific sustainable strategies, including green roofs
	Denver, CO	Green Buildings Ordinance	2018	New buildings with 25,000 ft^2 or larger must have reflective, light-colored roof coverings on the roof or a portion of the roof and must also incorporate one of the following measures. They must install green roofs or solar panels, achieve LEED gold, purchase renewable energy, or pay a per-square-foot fee The green roof ordinance was initially passed in 2017 but was revised in 2018 in response to challenges associated with the green roof mandate
	Seattle, WA	Seattle Green Factor	2006	A score-based code requirement aims to enhance the quantity and quality of landscaping in new developments. Projects within specified zoning districts must attain a minimum score by earning credits tied to diverse green landscaping strategies, including green roofs
Rebates, tax abatements and refunds	Hamburg, Germany	Green Roof Strategy	2014	Building owners who install green roofs may receive subsidies covering 30–60% of the costs associated with voluntary greening measures. The city plans to enforce mandatory green roofs and green façades for all suitable buildings through legislation

(continued)

Table 1.3 (continued)

	City	Law/policy title	Year enacted	Operative provision
	New York, New York	Green Roof Property Tax Abatement Program	2008	The program offers a one-year tax abatement for the installation of green roofs on residential and commercial buildings in New York City
	Palo Alto, CA	Green Roof Rebate	2008	Provides a rebate of $1.50/ft^2 for installed green roofs, with a limit of $1000 for residential properties and $10,000 for commercial properties. This rebate is offered as part of the city's Stormwater Measure Rebate Program
Zone interactions	Austin, TX	Green Roof Density Bonus	–	Green roof installations can earn up to 1 ft^2 of bonus floor area for 1/ft^2 of planted bed provided on a vegetated roof
	Chicago, IL	FAR Bonus	2015	A bonus of up to 2 floor area ratio (FAR) is awarded for downtown developments with green roofs covering more than 50% or 2000 ft^2 of the roof area
	Philadelphia, PA	FAR Bonus	–	An FAR bonus is awarded to residential developments with green roofs and to "green buildings" that qualify for the designation through the installation of green roofs
	Portland, OR	Ecoroof FAR Bonus	2009	An FAR bonus is granted for residential developments within the Central City Plan District. Developers may earn one additional square foot for 10–30% green roof coverage, two additional square feet for 30–60% coverage, and three additional square feet for coverage exceeding 60%

Singapore's City in Nature vision, a key pillar of the Singapore Green Plan 2030, promotes skyrise greenery. The Singapore Skyrise Greenery Incentive Scheme (SGIS) provides financial incentives for building owners and developers to install green roofs and vertical greenery. SGIS covers up to 50% of installation costs, with funding caps of $200/m^2 for rooftops and $500/m^2 for vertical greenery. The program supported over 200 buildings, including edible gardens, recreational rooftop gardens, green roofs, and green walls [49].

Seoul's Roof Garden Support Project, which began in 2002, created 661 roof gardens totaling 278,242 m^2 by 2013, at a cost of KRW 59,338 million [50]. The project produced several benefits. It reduced building surface temperatures by 60–70%, lowering energy use. It provided ecological spaces and recreational areas for citizens. It also served as an educational resource on the environment. Seoul has tried to improve roof gardening by revising the Building Act and securing dedicated funding. The city received KRW 675 million from a green fund competition to create roof gardens on six public buildings totaling 2090 m^2. This external recognition validated the ecological benefits of Seoul's "Creation of 100,000 Green Roofs" campaign. The city collaborated with various organizations to facilitate the roof gardening initiative.

The City of Paris and Paris Climate Agency have enhanced advisory services for energy savings. Paris implemented a Climate Plan in 2018, focusing on the energy dimension of urban development and finding innovative solutions. The plan aims to reduce emissions by 25% by 2020 and 75% by 2050 through a transition to low-carbon food and energy systems, benefiting public health, climate protection, jobs creation, and living conditions. In addition, Paris aims to achieve 40% permeable green space to adapt to the effects of climate change. To build resilience by 2050, Paris plans to enhance green spaces, tree coverage, ground permeability, and surface materials, while improving urban surfaces and underground spaces [51].

The Melbourne Urban Forest Fund aims to establish public-private collaborations to expand the city's green spaces. As the population grows and climate changes, increased green cover provides environmental, social, and economic benefits beyond the City Council's current efforts. The fund provides financial incentives for integrating new green spaces, offering a dollar-for-dollar matching grant. Grants range from $25,000 to $500,000, with a minimum total project value of $50,000, requiring an applicant contribution of at least $25,000 and an additional match fund of $25,000. To receive the maximum $500,000 grant, the applicant must invest at least $500,000 [52].

Figure 1.1 shows a comparison of different cities for green roof area per capita. Basel has the largest public green roof access, followed by Stuttgart. Tokyo and Singapore have smaller green roof areas per capita due to their high population density.

1.2.3 Social Sustainability of Tall Buildings

We live in an era in which the construction of tall buildings is booming. In urban centers today, high-rise office and apartment buildings are being promoted as a necessity for economic prosperity, even though there are fears that such mass housing development creates social gentrification issues and may not offer appropriate urban experiences. High-rise towers including high-rise apartments, have been dominant

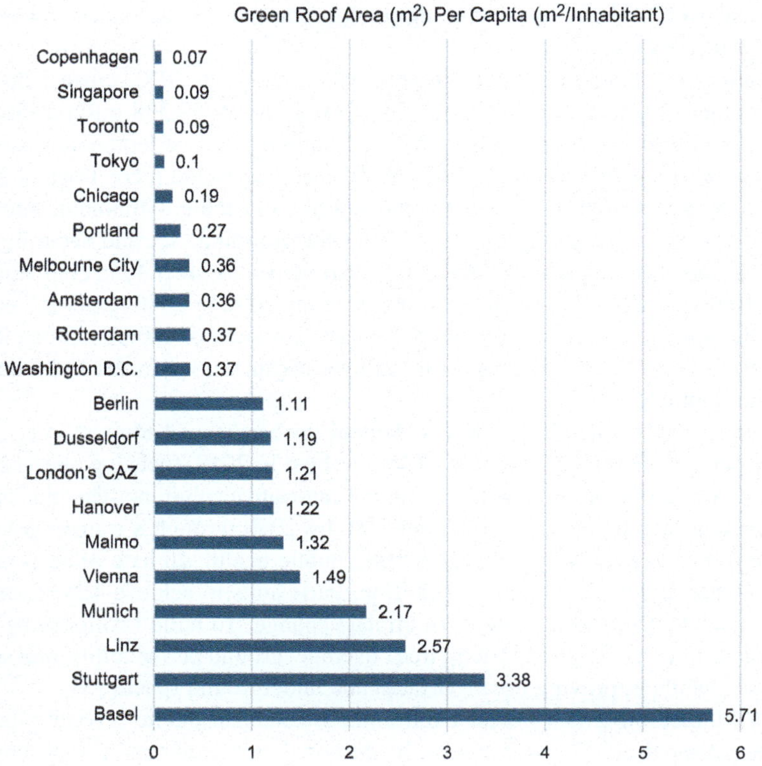

Fig. 1.1 Green roof area per capita comparisons for different cities

in redevelopment activities in both developed and rapidly growing cities. Flagship projects that raise the new economy image are one of the redevelopment tactics of these cities, as part of a broader goal to bring in capital and stimulate local prosperity.

Some of developments involve the redevelopment of brownfield sites, for example—an obsolete industrial railway yards as well as waterfront redevelopment areas. In their aspiration to send a "high-end" market positioning message to target audiences, these buildings are often described as "iconic architecture". This current development trend has seen high-rise apartments proposed and constructed primarily for profit, often at the expense of broader social good. In the age of globalization, high-rise residential and commercial buildings have become part of crafting a skyline image of cities into the twenty-first century. Asian cities become the global hubs of dynamic and high-density urban living.

The tendency in the rehabilitation of older social housing has increasingly shifted toward the demolition of existing buildings instead of their renovation. Such an approach is intended to result in better social and health returns by relocating people from aging housing blocks to newer houses with improved amenities. Nonetheless, the practice of tearing down and replacing social housing can generate additional problems, including community displacement and gentrification. Research examined the gentrification of luxury real estate growth in New York City by analyzing 943 projects built between 2000 and 2020. New luxury developments increased building height by an average 6.8 stories and more than doubled building volumes compared to neighboring structures, leading to middle-class displacement and super-gentrification in already gentrified areas [53]. Another study also examined how environmental goals aimed at increasing density and promoting sustainable housing near transit-oriented development (TOD) can overlook social justice concerns and contribute to gentrification [54].

It has been studied how high-rise buildings affect public safety and criminal activity, as well as how centralized rational planning influenced the urban dynamics. The vast use of superblock zoning as well as the elimination of narrow street networks and mixed-use neighborhoods diminish pedestrian activity, social interactions, and visual surveillance [55]. Research has shown that the building height in New York City has an impact on crime rate, vandalism activities and destruction [56]. Other environmental design factors, such as block size, occupancy types, open spaces, and sidewalk networks, can enhance public safety, whereas low surveillance opportunities can create safety issues.

Tall buildings can adversely affect the mental well-being of residents. Residents in tall towers are more vulnerable to depression and isolation than those living in low-rise flats [57]. The fundamental problem is loneliness because high-rise living is associated with weaker neighborhood and social networks. A shortage of social facilities within towers, high occupant turnover, degraded external areas, and uncontrolled interaction among residents can create an environment conducive to detachment and negative health outcomes among households in high-rise buildings. Recommendations include the provision of accessible semi-public amenity, with spacious, well-lit corridors and shared entrances, contributes to social cohesion [58]. Additionally, carefully designing the interfaces of private and semi-public spaces can increase intervisibility and give residents a greater sense of control over their environment [59].

Connectivity: High-rise buildings accommodate larger economic activities and are effective in space management, particularly in urban centers where space is limited [60]. Tall towers with mixed uses of residential, business, retail, and recreation, enhance social interaction, activate street life, and create vibrant and inclusive communities [61]. High-rise buildings not only provide social spaces but also improve look social connectivity by living walls, sky gardens, and observation decks [62].

The development of tall buildings in transit-oriented developments (TODs) areas provides access to necessary services, employment opportunities, and other facilities. Efficient development patterns promote increased walking, a reduction in vehicular emissions, and dependence on personal automobile. This is advantageous to residents, as it promotes social equity and accessibility. However, potential challenges include high levels of noise, public transit vibrations, and density-related crime [63]. The environmental benefits and public transportation access from TOD are very important, and these benefits can outweigh negative impacts through careful planning, urban design, and community participation. TOD development profoundly impacts marginalized communities and shapes neighborhoods. Development must balance equity concerns, prevent unjust displacement due to gentrification, and ensure that affordable housing and cultural heritage are preserved as part of the development process [64].

Resilience: The occupants of tall buildings can be protected through resilience-enhancing features that help them recover from natural hazards such as typhoons and earthquakes. In emergencies, these buildings can serve as shelters, accommodating large numbers of people due to their large, well-equipped spaces. Additionally, community-oriented design strategies enhance community resilience through redundancy, strong envelope barriers that prevent interruptions to vital services, effective evacuation planning, and community programs that promote safety and social sustainability.

As an example, to provide an additional level of protection to the occupants, the stairways within the Burj Khalifa are constructed of fire-resistant concrete. Since it is not possible to evacuate all occupants down 160 floors during a fire, the structure is provided with refuge area at every after 25 floors. These areas are provided with fire-resistant walls rated for a minimum of two hours and are designed to prevent smoke infiltration, as the refugees and stairwells are internally pressurized. These requirements suggest a "defend-in-place" strategy, allowing a certain level of protection during both actual emergencies and perceived threats, while safeguarding exit routes or providing safe refuge areas as practically possible [65].

The tower also features some technologies that are becoming prerequisites for the next generation of highly resilient supertall buildings. The structure of the Shanghai Tower incorporates the concept of vertical cities, addressing the challenges imposed by the site, including exposure to high wind, seismic activity, and clay-based soil. The central concrete core forms the backbone of the structural system. To support the lateral movement of each vertical neighborhood, the double-belt trusses of the steel outriggers and super columns work in tandem with the structural core. The building is vertically zoned by fire refuge and mechanical floors, which extend outward at each zone, further enhancing safety and structural performance [66].

Chapter 2
Sustainable Design Principles

Abstract This chapter investigates the sustainable design principles of building enclosures for tall buildings located in hot climates. Building enclosures are the primary building systems that mediate the relationship between the indoor and outdoor environments. Building enclosures play an important role in a building's energy operational demand, occupants' comfort, and architectural expression through the control of air leakage, moisture, thermal performance, solar radiation, and daylighting. Bioclimatic façade strategies are central to sustainable tall building enclosures, enabled by climate- and site-responsive design approaches that minimize reliance on mechanical conditioning by using local and renewable resources, such as wind, daylight, water, and vegetation. In hot and humid climates, the integration of shading devices, operable openings, and vegetation-based systems helps moderate microclimates, enhance user comfort, and reduce energy demand for space conditioning. The façade design also supports building program requirements, and enclosure strategies must address public identity, spatial complexity, and diverse functions at both inter-building and intra-building scales. Tall building enclosures typically consist of storefront, curtainwall, window wall, and opaque wall system assemblies. Building enclosure materials commonly include metals, glass and ceramics, wood, and composites. Material selection and system assemblies meet essential performance requirements such as structural strength, life safety, waterproofing, and reduced environmental footprint. Enhanced environmental sustainability is achieved through the implementation of high-performance environmental control systems and bioclimatic strategies such as natural ventilation, shading, daylight, and controlled solar gain. Further discussion addresses the integration of renewable energy systems, such as building-integrated photovoltaics (BIPV), with focus on available technologies, performance challenges, and end-of-life considerations. Nature-integrated building enclosures further enhance environmental and social benefits through biophilic design, biodiversity support, and pollution mitigation. Overall, S = sustainable tall building enclosures contribute positively by strengthening connections between people, the built environment, and nature.

K. H. Kim, *Continuity in Sustainable Facades*,
https://doi.org/10.1007/978-3-032-19841-9_2

Keywords Sustainable tall building enclosures · Bioclimatic façade design · Building-integrated photovoltaics (BIPV) · Biophilic and nature-integrated design · Occupant health · Well-being

2.1 Sustainable Facade Design Principles

Sustainable façade design is one of the essential architectural elements that contributes to building functionality, longevity, and aesthetics. Sustainable façade design balances multiple design requirements, including climatic and site specificity, programmatic and spatial considerations, user response, cultural and social connectivity, and local materials and construction techniques. As an interface between indoor and outdoor environments, building enclosures offer technical functions such as air leakage control, thermal regulation, moisture control, shading efficacy, and energy efficiency.

2.1.1 Bioclimatic Facades

Sustainable façade design considers the local climate, site context, topography, and vegetation to maximize available natural resources and reduce the reliance on active building systems, such as air conditioning or mechanical ventilation equipment. Focusing on climate and site-specific design ensures that the building envelope utilizes local recourses such as light, wind, water, and vegetation and reduces the reliance on nonrenewable resources.

Climate- and site-responsive, or bioclimate responsive facades play an important role in achieving ecological sustainability through shading, natural ventilation, winter heating, and daylighting. Since facades are the interface between the outdoor and indoor conditions, they are the main building element which harnesses wind for natural ventilation, daylight, and solar shading strategies. Accordingly, user satisfaction is enhanced because they have opportunities to interact with and control such environmental conditions. Bioclimatic design enables buildings to better adapt to their local climate, while simultaneously improving occupant comfort and reducing energy consumption.

The high humidity and elevated temperatures of hot climates pose significant challenges to design comfortable and energy-efficient buildings. However, with the appropriate façade design utilizing bioclimatic design principles can lead to cost-effective energy efficient buildings. Utilizing green roofs and living walls as well as incorporating water features help shading and cooler microclimates. Implementing solar energy technologies can power the building operation. By understanding the local climate and conditions, building enclosures can leverage bioclimatic strategies that are environmentally friendly, healthy, and aesthetically appealing.

Space and Program Considerations: Facade design addresses both functionality and aesthetics, representing architectural space and programs. In other words, the spatial programs of a building should inform façade design and in turn, the façade design should support and facilitate programmatic requirements. Whether functional or aesthetic, façade design portrays building programs and underlying structural systems and materials. Whether a space is intended for work environments, communal areas, lobbies, or kitchens, these unique façade designs can provide optimal environmental conditions to accommodate diverse programmatic needs. Innovative and context-sensitive facade designs can provide positive user experience and productivity, while providing effective functional solutions for a wide range of spatial and programmatic requirements.

The façade design of the Universidad EAN, located in Colombia is a good example of how facade design takes into consideration of the programmatic needs of a space. The design includes a visual hierarchy that distinguishes different levels but also focuses on how the façade best serves the learning environments and public spaces. The facade is designed with solar brise-soleil elements made of recycled metals. The main façade features large three-dimensional screens that provide both functional and aesthetic support for the program behind it. The design is fully integrated into the community, and architectural elements such as large apertures help provide natural light to interior and open spaces.

The facade is a key architectural element that conveys compelling narratives about the site, building, space, users, history, and culture. Depending on the spatial organization, certain elements such as stairs or balconies can be implemented to give distinct identity to façade designs. Other elements used in façade designs include terraces or sky gardens, which are strategically planned throughout the building to promote community building and social mixing. Façade features can also be used as branding tools to express values and promote a business's identity using color, texture, shape, and other design elements. These features can also create a memorable experience for visitors and can be used as a social catalyst that enhances interaction among occupants. Incorporating balconies, atriums, sky gardens, and terraces can serve as social features that encourage encounters and engagement among users.

Sustainable Façade Technologies: Tall tower enclosures can consist of curtainwall, window wall, and storefront system depending on how the structural loads are transferred and how the enclosure is set out relative to the primary structure. The enclosure system can consist of glazing facades, opaque walls, or a combination of both. The curtainwall system is suspended from the slab and installed in front of the slab. The window wall system, on the other hand, is typically installed between slabs or resting on the primary structure. The storefront is similar to the window wall in terms of structural load path, but it refers specifically to enclosures installed at the ground level, where typically, retail spaces are typically located (Fig. 2.1). The curtainwall is composed of framing members and infill panels. The infill panel within the frame consists of vision glass and spandrel panels. Vision glass allows for greater daylight penetration and view-out to the city, while the spandrel panel conceals structural and mechanical components at the floor level. Spandrel panels also contribute to energy efficiency and fire safety performance.

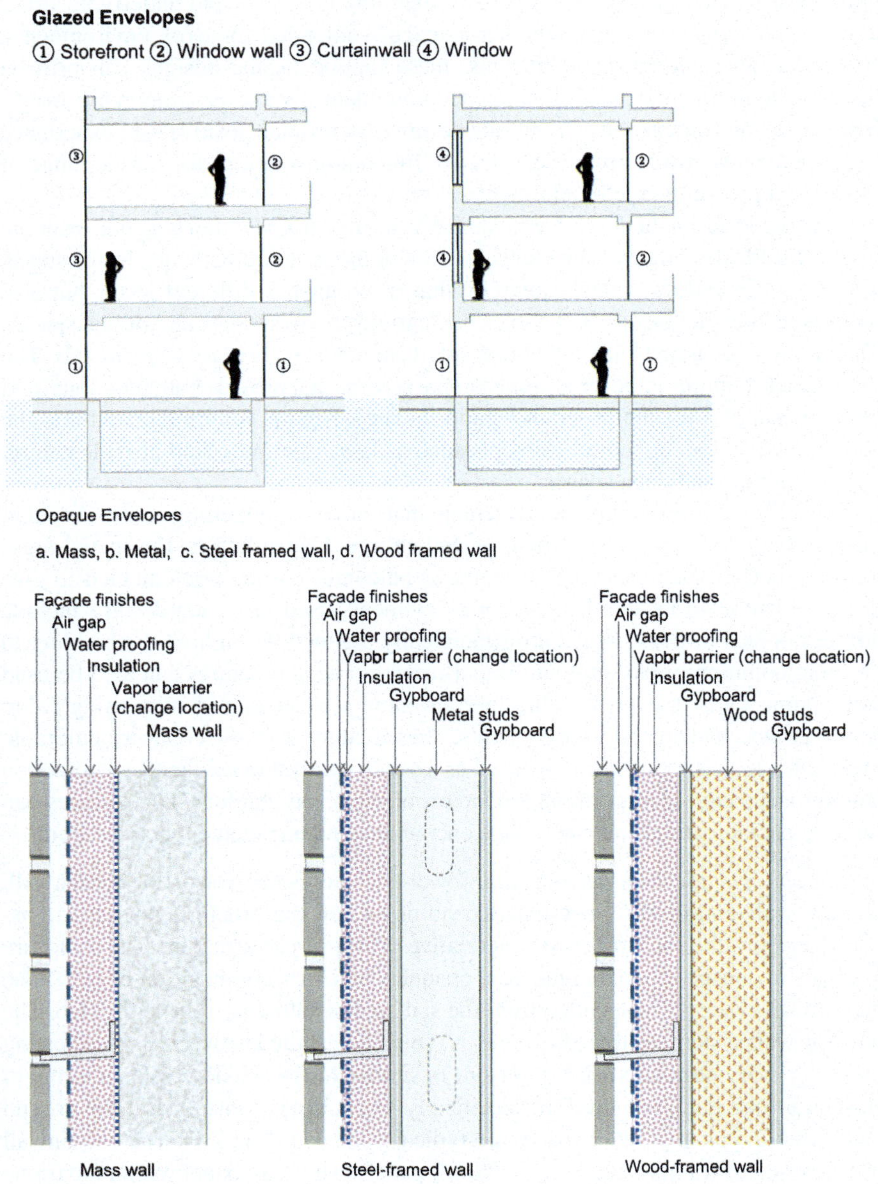

Fig. 2.1 Various façade types for glazing facades and opaque wall construction

Opaque walls provide enhanced thermal protection and privacy. They can be divided into mass wall and lightweight construction, based on the structural backup wall assembly. Mass wall systems are typically constructed using concrete masonry units (CMUs) or precast concrete panels, and include layers such as sheathing, vapor barrier, insulation, waterproofing membrane, air cavity, and final cladding. Lightweight construction uses wood-framed or steel-framed walls, which also incorporate sheathing, insulation, vapor barrier, waterproofing membrane, air cavity and final cladding. The vapor barrier is typically installed on the warmer side (inner surface) of the insulation layer, and the layering sequence in the figure is based on heating-dominant regions. In cooling-dominant regions, both the vapor barrier and water proofing membrane are positioned on the colder side (outer surface) of the insulation. A well-designed building enclosure system that balances structural integrity, energy efficiency, life safety, cost-effectiveness, durability, occupant well-being, sustainability, and aesthetics is essential for creating high-performance, environmentally responsible, and socially conscious buildings. Figure 2.1 explains the glazing wall and opaque wall construction types.

Tall building enclosures provide functional and symbolic design. The symbolic nature becomes a strong aesthetic expression of a certain architectural style or cultural consideration, giving a building its identity. Generally, façades are divided into two types depending on transparency and opaqueness: The first type of façade uses glazed enclosures to create light and views, enhancing the perception of interior spaces and their connection to the ever-changing outdoor environment. These are mostly seen as windows, clerestory, or skylights. The second type focuses on the opacity of architectural walls, influencing solidarity, controlling the public image, or responding to space organization such as servant programs or public services. For example, the storefront façades and the atrium enclosures of the World Trade Center Museum Pavilion are glazed, which has a positive impact on their site, history, and functionality. They feature large glass facades integrated with metal façade frames that give them a distinctive architectural identity. The reflectivity of glass materials has a symbolic meaning of memory and reflection of the past. They also have additional other characteristics such as punched windows that create natural light during the day and dynamic lighting effects at night.

By taking advantage of advanced materials and technologies, tall tower enclosures can be designed to match any design intent or color schemes. The wide range of material pallets allows design teams to engage in diverse architectural features, while promoting environmental stewardship and social connectivity. With advancement in construction technologies, facade materials now span a wide variety, ranging from metals and glass to polymers, wood, terracotta, and composite components. Material selection can be guided based on physical performance, longevity, and ecological footprint such as embodied carbon ($kgCO_2/kg$) or embodied energy (MJ/kg).

Sustainable Materials: A building façade is designed using key design science principles, influencing material selection as well as fabrication and construction techniques. Facade materials can be categorized into four primary groups: metals, glass and ceramics, wood, and composites. Materiality is closely connected to the physical characteristics of building enclosures and architectural elements, including façade performance, aesthetics, longevity, and sustainable attributes. Functional requirements and aesthetic goals differentiate facades in material type, composition, and scale. These materials help protect building enclosures from collapsing by fire, wind, rain and snow while also expressing their building types and local cultural identity.

Material choices can express coherence and functionality. For instance, stone or brick facades may reflect neighborhood identity while wood cladding can soften the visual character of surrounding buildings without compromising durability. Metal and glass facades reinforce contemporary architectural images. Sustainable façade design adopts appropriate materials and assembly techniques in response to local identity, climatic conditions of the site, spatial programmatic requirements, visual appealing, indoor environment quality, and other functional considerations.

2.1.2 Energy Efficient Building Systems Integration

Bioclimatic design strategies discussed in the previous section should consider the integration of energy efficient building systems to further enhance overall building performance. Once bioclimatic design strategies are implemented, operational energy demand can be further reduced through the use of energy-efficient active environmental control systems. Energy-efficient environmental control systems—such as air conditioning, space heating, ventilation, and lighting—can be effectively integrated with building enclosures to enhance overall system performance rather than operate as isolated systems.

Facades as a Ventilation System: Building enclosures are vital for natural ventilation, which uses air flow and thermal buoyancy effect to circulate indoor air without relying on mechanical systems. Proper façade design and window orientation can channel winds into the building. Strategic window placement promotes cross-ventilation and stack effect: high-point outlets release hot air, while low-point inlets draw in cool air. Adjustable casement or louvred windows control airflow direction and volume. The efficiency of different natural ventilation strategies can be assessed using the computer simulations of the airflow and the openings. The research findings showed that the number and size of the openings have a significant effect on the airflow and pressure profiles, with excessive ventilation openings negating the benefits of the stack effect [67]. The stack effect and natural ventilation increase in spaces with significant height differences between open inlets at lower levels and open outlets at higher levels. This promotes air movement caused by convection and buoyancy.

Window opening sizes and locations significantly affect ventilation rates and indoor airflow patterns. When cross-ventilation (CV) flows are considered, two openings positioned on the opposite side of a room perform better than two openings placed on the same facade location. CV flows increase as the area of the leeward aperture increases. Facade modifications, such as louver blinds and eaves, alter the CV airflow in buildings. In the case of louvers, solar heat gain from the sun's position can be reduced, which also affects CV flow patterns. Research examined changes in louver angles and their relationship with indoor airflow velocity, direction, and ventilation rates [68]. Lower louver angles resulted in higher indoor air velocities, whereas higher louver angles affected airflow direction and reduced inlet velocities.

Architects and engineers encounter challenges, especially when configuring natural ventilation openings within building forms and layouts. Indoor air exchange rates and the airflow velocity are largely determined by the size and location of ventilation openings. A Texas-based study investigated a single-family house with a gable roof in plan mearing approximately 13.2 m wide, 8.3 m deep and 5 m high [69]. The findings indicated that the cross-ventilation rates increased with wind velocity component perpendicular to the openings. The Refresh Cube Campaign is a full-scale experimental setup designed to investigate wind flow characteristics and ventilation performance [70]. It focused on sheltered and exposed conditions using isolated cubes to simulate effect of urban buildings. Result indicated that the ventilation rates of a building in sheltered settings were reduced by 50–90% compared to the building without surrounding buildings. The work emphasized that collected site specific wind speed data are crucial for predicting the cross-ventilation performance in urban environments, and demonstrated the lack of simplified urban ventilation model.

Double skin façades (DSFs) have gained popularity in office buildings in Europe, Asia, Australia, and North America for their impact on indoor environment and indoor and outdoor space interaction. DSFs offer energy efficiency and user experience by improving ventilation, daylighting, acoustic barrier, and better thermal comfort. Energy efficiency is enhanced by reducing heat transfer and regulating solar gains while maintaining outdoor views. Natural ventilation is achieved through the thermodynamic and aerodynamic interactions between the double skin façade and indoor space, which enhances indoor air quality and thermal comfort, even in extreme climate conditions. Wind pressure and building geometry strongly influence DSF airflow and, making holistic aerodynamic assessment essential. Computational models using computational fluid dynamic software are employed to examine how buoyancy and wind pressure enhance natural ventilation in DSFs. In cold climates, airtight construction of DSFs have been implemented in German office buildings to reduce the heat loss through uncontrolled air filtration [71]. Ventilated DSFs perform effectively in hot and subtropical climates, such as Hong Kong, where solar gains are reduced and natural ventilation is enhanced [72]. Even in hot climates, DSFs can remain practical and energy efficient, as integrated shading devices can effectively control solar gains and prevent overheating.

When occupants can control the amount of airflow and the temperature, their comfort level increases. Previous studies suggest that when the occupants have control over such variables, they report higher levels of satisfaction with respect to air quality, temperature and general comfort of the space [73]. Previous studies also indicate that offering personal control of environment factors increases morale and fosters a sense of ownership [74]. Occupants have greater agency, acting as active participants rather than passive receivers. This agency increases engagement, productivity, and occupant loyalty. Allowing the occupants to modify the amount of outdoor air intake and ventilation rates is therefore an effective method to enhance satisfaction and improve health and well-being.

Facades as a Daylighting System: Building envelopes are the primary architectural element for daylight penetration as well as view openings for the users. By allowing natural lighting to illuminate the space, electrical lighting loads are minimized, and energy efficiency is enhanced. The quality of views and the performance of view-out through windows support occupants' experience, their productivity, and well-being. Daylighting strategies utilize creative and functional facade design to enhance the aesthetics while also creating healthier indoor environments for occupants.

The architectural layout and the placement of daylight apertures across building perimeters play an important role in not only illuminating the space but also creating engaging spatial effects. In regions with higher solar angles, a daylight window near the ceiling, clearstory, and skylight would allow for a greater inflow of daylight. North-facing apertures allow diffuse daylight while ensuring minimal glare and solar heat gain. East- and west-facing windows have both advantages and disadvantages in that they can help control the circadian rhythm but may introduce glare and excessive solar gain during the sun rise and sun set times. In addition to window location, facade components such as window sills, light shelves, and even window materials can further enhance daylighting distribution into interior spaces.

Daylight metrics are advancing to better measure the interactions between users, the building, and ever-changing climate conditions. The New York Times Headquarter incorporates ceramic-coated external shading systems coupled with curtain-wall systems. Its energy and daylighting performance were validated through computer simulations and full-scale mock-up testing [75]. These technologies regulate the amount of daylight entering interior spaces, reduce glare, control heat gain, and increase effective thermal comfort. Simulation tools are becoming increasingly important for the validation and realization of sustainable facade design strategies, particularly for complex fenestration systems that offer high performance and visual appeal [76]. These tools also facilitate a wide range of parametric analysis to assess how various facade design parameters, facade orientations, and building forms affect daylighting illumination, visual comfort, and energy consumption.

Static facades do not incorporate moving parts, but their fixed systems provide a balanced performance across various metrics, including view-out, daylighting, and shading efficacy. In contrast, dynamic facades (DFs) have kinetic elements that move to changing environments. DFs integrated into a building could provide enhanced benefits, including improved energy efficiency, increased user satisfaction, and flexible aesthetic creativity. These systems open and close in response to solar intensity and temperature changes and automatically regulate the amount of energy flowing through the facade, resulting in energy savings from heating, cooling, and lighting energy demands [77]. Projects such as the Kiefer Technic Showroom and Al-Bahr Towers are how DFs have been implemented in real world operations and minimize the energy use from electrical lighting and HVAC systems. Several studies report that building energy and CO_2 emissions can be reduced by 20–30% by intelligent facades. There is also greater user interaction with DFs, which leads to increased satisfaction and productivity. Their adaptability makes them suitable for a wide range of building types and uses.

The quality of views framed by windows involves physical, aesthetic, and psychological experiences. In general, various view-quality metrics position windows as both functional and experiential elements in enhancing indoor environmental quality and occupant satisfaction. A study explored the relationship between windows, views, indoor environments, and user satisfaction levels [78]. The result indicated that windows positively influence users' perception in the spaciousness, brightness, and overall room and view satisfaction. Participants preferred more natural views to trees and other vegetations over built or artificial scenes. This means that windows with a scene of greenery can improve user satisfaction and the quality of the indoor space. Indoor scenes with a view out of the window towards vegetation are generally more pleasant and inviting.

Another study examined user comfort and behavior in office environments in relation to façade design, focusing on satisfaction with window and shading device control [79]. Four buildings with similar workspace configurations but different façade designs were compared, considering user control over ventilation and daylight, and the visual perception of transparent façade portions. The study concluded that user comfort tends to decrease in spaces that do not provide effective user controllability. The ability for occupants to control windows and shading devices results in higher satisfaction levels and greater tolerance of environmental discomfort. These findings emphasize the importance of considering not only comfort and energy efficiency aspects but also user controllability and visual appealing in facade design to improve user satisfaction.

High-quality window views are used in building design evaluation, performance assessments, and certification programs. For instance, the WELL Building Standard and LEED (Leadership in Energy and Environmental Design) incorporate view-related criteria in their assessment frameworks [80]. By measuring and proving high-quality user experience, architects, designers, and engineers can create built environments that enhance occupant satisfaction and well-being through effective window placement and quality views.

Facades as a Heating and Cooling System: Building enclosures significantly influence the thermal performance of tall buildings. Designing building facades with natural heating and cooling strategies in mind enables reduced dependence on mechanical heating and cooling systems, therefore resulting in lower energy consumption rate and improving sustainability.

Thermal insulation and the mitigation of cold bridging are important energy-related attributes of building enclosures, as they minimize heat transfer between indoor and outdoor environments. The thermal mass of building facades helps regulate indoor temperatures. For example, brick facades equipped with adequate insulation and appropriate construction details behave like thermal sponges, absorbing heat during the day and releasing it at night while maintaining an optimum level of indoor ambient conditions. In other words, interior temperatures remain stable because thermal mass walls absorb excess heat during the day, preventing overheating. The interior from becoming too hot, and releases stored heat at night, ensuring that indoor temperatures do not drop too low. In winter, the thermal mass facades collect solar heat from the sun to warm the space and gradually release that stored heat toward a later time releases that heat. The integration of thermal mass structures and thermal insulation enables energy savings and enhances thermal comfort.

An office building in Jacksonville, Florida was pre-cooled utilizing night-time and weekend ventilation and was found to decrease cooling energy demands by 18% [81]. The study indicated that increasing thermal mass from 21 to 201 kg/m^2 (4.3–41.2 lb./ft^2) reduced indoor temperatures by 1–2 °C (1.8–3.6 °F) and lowered potential energy use. Heavyweight buildings consume less cooling energy than lightweight ones, especially at 24 °C (75.2 °F) indoor temperature. Nighttime ventilation in desert climates can reduce cooling loads by 27–36%, with greater savings observed in heavyweight buildings. For optimal thermal mass, dense materials with wall thicknesses of 5–10 cm (2–4 in) are suitable for daily heat storage, while thickness of up to 25 cm (9.8 in) support longer thermal storage periods.

Another study explored natural cooling strategies in temperate climates as alternatives to air conditioning [82]. Using buildings from a 2015 "Green Energy" project, researchers found that thermal inertia improved comfort and reduced energy use. Lightweight wooden frames offered less thermal mass benefits compared to lightweight cellular concrete walls during hot summer periods. Concrete ground floors and cellular concrete walls, combined with night ventilation and external shading, kept indoor temperatures at 25–26 °C (77.0–78.8 °F) even during heat waves, eliminating the need for air conditioning in residential buildings. This approach lowered the cooling energy demand by 67–75%, highlighting the potential of passive cooling strategies through the integration of thermal mass, night ventilation, and external shading.

While opaque walls contribute primarily to thermal insulation and thermal mass, transparent facades are responsible for regulating solar gain entering buildings. Façade elements such as overhangs, louvers, shutters, and brise-soleils can block direct summer sunlight while allowing lower-angle winter sunlight to warm interior spaces. Automated shading systems can adjust to sun position and weather conditions, optimizing comfort and reducing cooling needs.

The design of façade systems can also control solar gain by capturing solar energy for winter heating during colder months. South-facing facades (in the northern hemisphere) with appropriately sized windows can maximize solar gains, thereby reducing the need for artificial heating. Low-emissivity (low-E) coatings on windows increase the insulation value by minimizing heat transfer while allowing sunlight to enter.

Reflective materials and cool roofs effectively reduce heat absorption in buildings. Reflective facades bounce off solar radiation, lowering indoor temperatures and cooling needs. Cool roofs with high reflectance and thermal emittance reflect sun rays, passively cooling buildings and reducing energy use and emissions. Together, these solutions offer a comprehensive approach to managing heat absorption and radiation, mitigating rising temperatures and urban heat effects. Studies on mitigating urban heat islands (UHIs) and improving urban comfort have focused on New York and Athens, using simulations to assess cooling impact in medium and high-rise neighborhoods [83]. Results showed street-level temperature reductions of 0.1–0.12 K, with effectiveness decreasing as building height increases. On average, a 0.1 increase in roof albedo led to a 0.2 K temperature decrease. UHIs result from impermeable surfaces and human-generated heat. Cool roofs reflect sunlight, reducing inside temperatures by 6–8 °C (42.8–46.4 °F) and roof surface temperatures by 40 °C (104 °F), lowering air conditioning demand. Green roofs offer additional benefits, including improved air quality, reduced stormwater runoff, and wildlife habitats. They can cool surrounding air and reduce indoor temperatures by 5–8 °C (41–46.4 °F) through evapotranspiration.

Green facades and roofs assist in creating a sustainable, eco-friendly built environment. They can provide additional insulation through additional thermal mass from soil or other medium, and biomass, which minimizes energy transfer through the green layers. Vegetation contributes to heat absorption and reflection and improves the insulation of buildings, potentially reducing energy bills and environmental emissions. They filter the air and other environmental pollutants, benefiting the urban ecosystem and users' health and wellbeing. Incorporating such nature-based solutions can contribute to sustainable solutions to modern urbanization problems and anthropogenic pollution by improving environmental performance and decreasing energy demand. Research on vertical vegetation systems (VGS) has identified three types: living wall systems, indirect green facades, and direct green facades [84]. A case study building with nine VGS cases was analyzed, focusing on microclimatic variations in temperature, humidity, and CO_2 levels. Results showed that VGS significantly reduced exterior wall surface temperatures compared to bare walls due to insulation, evapotranspiration, and shading. This approach improves indoor comfort, reduces cooling and heating energy use, and enhances occupant health. VGS facing indoor environments also help lower CO_2 levels and air pollution, regulate relative humidity, and enhance indoor air quality.

2.1.3 Renewable Energy Integration

The global power requirement is around 15 terawatts (TW) [85]. The sun's energy (173,000 TW) can supply this demand in under an hour. In 2019, PV solar energy production increased by 22%, accounting for 3% (730 TWh) of global electricity generation [86]. Prior to the 1970s, PV cells were used mainly for satellites and specialized space applications. Between 1965 and 1990, they served small-scale niche markets, leading to technological development and standardization. Since 1990, increased implementation and policy support have driven widespread market deployment, with the global solar PV market spreading across the Americas, Europe, and Asia [87]. Solar PV applications for the built environment include utility-scale power production and building-integrated PV (BIPV). BIPV integrates PV cells into building rooftops and facades, allowing on-site electricity generation for direct building use or surplus supply to the grid.

The first BIPV project was realized in Aachen, Germany in 1991, with Pilkington producing solar modules for BIPV applications in 1993 [88]. Environmental and technical factors affect solar cell efficiency and longevity, including dust, shading, and cloud cover. Partial shading can cause significant power loss and system damage. The high upfront cost is a barrier for stand-alone BIPV facades, but tall contemporary buildings often use fully glazed curtainwalls for aesthetics and sustainability. BIPV curtainwalls can offset costs through energy savings and workplace enhancements. BIPV curtainwalls offer multiple functionalities, leading to energy reductions and increased occupant satisfaction. Rooftops are generally preferred for PV modules, but as urban populations grow, BIPV curtainwalls will play a key role in net-zero energy retrofits. The prefabricated BIPV curtainwall allows for faster installation and improved quality control, helping to overcome challenges in dense urban settings.

Solar Cell Technology: There are three main solar cell technologies. Crystalline silicon accounts for 90% of the global market, with annual growth around 30% [89]. Monocrystalline and polycrystalline cells have maximum efficiencies of 19.8% and 20.4% respectively [90]. Thin-film technology uses less or no silicon and includes amorphous silicon, cadmium telluride, copper indium gallium selenide, and gallium arsenide. Efficiencies range from 5 to 18.8% [91]. Emerging technologies include hybrid cells (25.6% efficient), carbon nanotube cells (15% efficient), and dye-sensitized cells (15% efficient) [92].

Over the past five years, solar energy deployment has increased due to advancements in solar technology. While solar panels do not generate greenhouse gases during operation, they contain hazardous chemicals that make proper disposal challenging after their approximately 25-year lifespan [93]. This is a concern as standard PV panels are difficult to recycle due to the presence of lead, cadmium, silver, and other harmful materials [94]. PV panels can do more harm than good if the hazardous chemicals involved in production and installation are not carefully managed. Damaged panels have often ended up in landfills. There is evidence of possible solutions to the issue of disposing of solar panels. Some countries, like Australia, are reclaiming and reselling parts of used panels at reduced prices, making solar technology more affordable [95]. Europe has relatively high sunshine levels but comparatively low solar technology adoption. Europe has also passed the Waste Electrical and Electronic Equipment Directive (WEEE Directive) to improve solar panel recycling and end-of-life management [96]. Proposing and enforcing laws to regulate PV panel disposal is important as installations increase, ensuring responsible end-of-life practices.

The building sector is expected to experience a significant increase in both energy consumption and pollutant emissions due to the rising demand to accommodate growing populations. Consequently, incorporating sustainable strategies and integrating solar energy systems in building enclosures has emerged as a pivotal approach to curbing fossil fuel use and enhancing the energy independence of both existing and new buildings. Research was conducted to improve the energy performance of apartment renovations using bioclimatic design, system updates, and the integration of double skin facades with solar cells, taking into account building location, tilted angles, and facade orientations using a simulation tool. The results indicate that the renovated case study building, integrated with a photovoltaic system, contributes to offsetting 63% of the total energy use [97]. Solar integration is a key strategy to achieve net-zero energy buildings. Apple's new HQ in Cupertino is North America's largest LEED Platinum-certified net-zero energy office building, powered by 100% renewable energy from various sources, including rooftop photovoltaic (17 MW) and biogas fuel cells, while feeding clean energy back to the grid during low-occupancy periods [98]. Building enclosures in urban environments could serve as prime locations for integrating photovoltaic systems because the rooftop area is relatively small compared to vertical surfaces. The FKI building enclosures in Korea consist of photovoltaic spandrel panels facing the sun and viewing windows facing the ground, designed for both maximum power production and aesthetic appeal. The power production from the photovoltaic facades and roof of the FKI contributes to approximately 3% of the total building's energy usage, which can power 274 homes and reduce 394 tons of emissions per year [99].

Table 2.1 shows potential BIPV solutions in real-world deployment. These include opaque cladding and transparent glass applications. The integration of BIPV with glass modules acts like a shading device and makes it possible to minimize solar gain. On the other hand, views and daylight penetration may be reduced.

2.1.4 Nature Integrated Facades

Green features have been an integral part of life for centuries, providing people with an intimate connection to nature. From ivy-covered walls to window boxes overflowing with flowers, green features add ecological benefits and aesthetic appeal. As we become increasingly aware of our environmental impacts, building enclosures are integrating vegetation to create sustainable structures that combine biophilic aesthetics and functional qualities. By incorporating vegetation into building enclosures, living walls can provide shade, insulation, biodiversity, phytoremediation, and even food. Living walls are usually composed of plants, soil, and water, and the vertical surfaces have living plant matter growing on them in some capacity. The plants absorb CO_2 and emit O_2 via photosynthesis, filter rainwater for use by natural vegetation in the surrounding environment and provide insulation from cold winter winds while trapping heat from the sun's rays during winter months. Depending on the design of the biofacades, they can also provide wildlife habitat, food, or visual privacy when desired. Biofacades provide biophilic benefits for users when interacting with them.

Although nature contributes positively to health and happiness, many people in high-rise buildings have limited access to nature. Plants affect psychological restoration and stress mitigation. Office workers working in rooms with plants reported 10% greater thermal comfort than in a control setting without plants [100]. Plants were found to increase human comfort and satisfaction in an office environment using subjective questionnaires and physiological measurements. Most people prefer

Table 2.1 BIPV typologies for vision and opaque wall applications and responding performance characteristics

BIPV types	PV integration with building enclosures		Performance characteristics/ differentiators
BIPV wall	Without back side ventilation	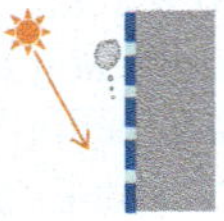	• PV cell heat up • Dust accumulation • Harsh environmental loading (i.e., rain, sunlight, wind) • No maximum power output due to cosine loss • No view-out/no daylight
	With back side ventilation	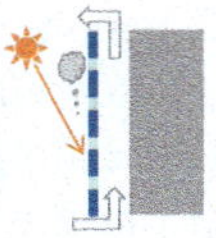	• Less cell heat up • Dust accumulation • Harsh environmental loading • No maximum power output • No view-out/no daylight
BIPV window	PV integrated with a window without backside ventilation	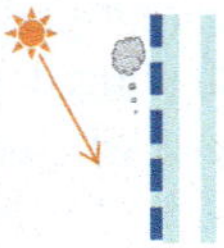	• Cell heat up • Dust accumulation • Harsh environmental loading • Permanent shading (i.e., no winter sun penetration) • No maximum power output • Little view-out/little daylight penetration
	PV as a permanent shading device with backside ventilation	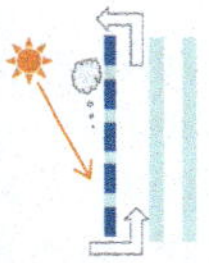	• Less cell heat up • Dust accumulation • Harsh environmental loading • Permanent shading • Little daylight penetration • Little view-out
	PV as a permanent shading device (static or dynamic) with backside ventilation	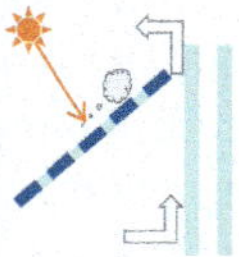	• Less cell heat up • Dust accumulation • Harsh environmental loading • Permanent shading • More view-out
Closed cavity façade BIPV (CCF-BIPV)	Encapsulation of regenerative solar modules inside of a closed air cavity ventilation		• Cells are cooled • No moisture buildup • No dust accumulation • Optimized shading and daylighting • Winter sunlight penetration • View out

window views, and green spaces viewed through windows show a positive association with improvements in human health and wellbeing [101]. Viewing green spaces from tall buildings also positively affects the physiological and psychological well-being of residents [102]. Through interactive processes, the public can learn more about nature and their benefits to the community and the environment. Their green color also provides psychological comfort and physiological regulation.

Vertical Landscape: Urbanization is reducing opportunities to interact with nature. With limited ground-level green space, vertical landscapes like green walls, balconies, and sky gardens can provide benefits such as air purification, noise reduction, and improved aesthetics, making cities more attractive places to live and work. Green walls are made up of plants and other vegetation that are grown vertically on the sides of buildings and can create interesting social pods and green spaces in cities. Vertical green walls can serve as a secondary biological membrane separated by a maintenance catwalk from the primary building skin. Greenery placed on intermediate floors or balconies of buildings adds an additional layer of aesthetics and serves as a communal space. By introducing green spaces on the upper levels of buildings, sky gardens create a buffer zone that not only helps reduce air pollution but also brings nature closer to users and visitors. These gardens can also improve rainwater collection and surface water runoff and provide water features for building inhabitants.

Incorporating nature into tall towers requires consideration of practical issues such as cost efficiency, quick returns on investment, operation, and maintenance. Automation can help make these green walls easier to maintain by taking care of watering, fertilizing, and maintenance processes. Automated systems can also monitor the health of the plants and provide alerts when maintenance is needed. The advancement of the internet of things (IoT) makes it possible for tall tower green walls to be installed in more places than ever before. Automating skyscraper biofacades has become a popular way to reduce energy costs, water usage, and carbon footprints while creating habitats for wildlife such as local species and biodiversity.

Numerous scientific studies demonstrate an innate connection between human beings and nature. Interactions with the natural environment can positively affect mental and physical health. This phenomenon is termed biophilia. The term "biophilia" was coined by social psychologist Erich Fromm (1964) and popularized by biologist Edward O. Wilson in his book The Biophilia Hypothesis (1981). Biophilia describes the innate human attraction to nature. This innate behavior has been defined as the love of life and its natural forms worldwide. Roger Ulrich's Stress Reduction Theory is associated with the restorative effects of nature on well-being [103]. His research, including a 1984 study on recovery from surgery, showed that natural views can reduce stress and improve healing outcomes. While Wilson introduced the biophilia concept, Ulrich's theory provided empirical evidence for how natural environments alleviate stress and improve health. Both theories have significantly influenced environmental psychology, architecture, and urban planning, emphasizing the importance of natural elements in built environments. Increasing interaction with greenery through building enclosures interventions may enhance positive attitudes toward nature and a willingness to protect the environment.

From improved environmental quality to enhanced productivity, biofacades can provide a range of health and well-being benefits to users. Even indirect interaction with nature can have a positive impact on mood. From virtual reality to watching nature documentaries on TV, studies have shown that these passive forms of interaction still create positive effects on people, including a sense of calm and relaxation [104]. Studies have shown that indoor greenery experiences in these environments can reduce stress levels and improve concentration in higher work performance. In classroom settings, incorporating greenery experiences into study areas can be especially beneficial for the mental health of university students who are dealing with the pressures of academic life [105]. Introducing plants has been proven to contribute positively to mental well-being, especially in areas with reduced space availability due to health issues [106].

Indoor plants have a range of benefits including improved air quality, natural attractions, and stress relief. They are especially beneficial for those with limited space or light while promoting mental well-being. Recent studies have shown that studying in a library room with natural environments, such as natural light and plants, can lead to higher cognitive performance than a control room without these features [107]. Access to green features potentially helps students stay focused and reduce stress levels while they study. More educational environments are recognizing the benefits of green spaces and investing in them, leading to a healthier and more sustainable environment.

Living in a high-rise building with access to nature is also beneficial for mental well-being. Recent research has revealed that having a connection with nature from tall buildings can enhance a stronger sense of connection to their communities and, most importantly, elevate their physical and mental health [108]. Therefore, it is essential for high-rise residences to provide windows with views onto natural landscapes. Some of these opportunities could be in the form of a central courtyard or garden space, as well as providing close proximity to a green wall or green balcony. Green features provide visual interaction and making connections to nature through both active and passive interaction, either indoors or outdoors, can mitigate mental stress and improve cognitive performance.

Bio-facades can be implemented using soil-based or hydroponic systems (Table 2.2). Soil-based vertical farming uses traditional soil-based planting methods, with plants grown in soil-filled containers on a building's exterior. This utilizes soil's natural growing medium to provide nutrients and support for plant health. Hydroponic vertical farming, in contrast, cultivates plants in nutrient-enriched water without soil. This soilless approach offers benefits such as improved space use, better growth control, and the avoidance of soil-borne issues. Both systems require additional structural support to accommodate soil, water, and plant weight as well as irrigation systems and other additional harvesting activities such as nutrient dispensers or maintenance loads.

Depending on construction and growing techniques, biofacades can be categorized into modular systems, wall systems, and hydroponic systems. Modular systems,

Table 2.2 Comparative analysis between soil-based biofacades [109, 110] and hydroponic-based biofacades [111–114]

Aspect	Soil-based biofacades	Hydroponic-based biofacades
Growing medium	Uses conventional soil-based growing media in planter boxes or pots affixed to the façade, such as potting soil or compost	Instead of using conventional soil, plants are cultivated using a nutrient-rich water solution
Plant selection	Maintains a large variety of soil-based plant types	Limited to hydroponically grown plants with low-maintenance varieties
Watering requirements	Requires frequent irrigation to keep soil moisture levels stable	Compared to traditional watering methods, automated irrigation systems offer an effective and environmentally friendly method of cultivating plants while consuming less water
Nutrient delivery	Relies on nutrients present in the soil or supplemented through fertilizers	Nutrients are supplied directly to plants through the water solution, ensuring optimal uptake
Weight considerations	Heavier due to the use of soil-filled containers or planter boxes	Lighter weight due to the absence of soil, reducing structural load requirements
Maintenance	Requires regular monitoring of soil moisture levels, nutrient content, and occasional fertilization	Requires regular monitoring and periodic flushing or replenishment of the nutrient solution
Aesthetic appearance	Natural appearance with soil and organic matter, suitable for traditional or rustic design aesthetics	Modern and streamlined appearance with clean lines, suitable for contemporary architectural styles
Adaptability	Suitable for outdoor and indoor applications, supporting a wide variety of plant species	Ideal for indoor or controlled environment applications, with limited plant selection compared to soil-based systems
Installation complexity	Relatively straightforward installation process involving containers or planter boxes attached to the façade	Requires careful installation of hydroponic panels or modules and integration of irrigation and nutrient delivery systems
Environmental sustainability	Can support a variety of microbial and plant communities and enhance soil health by adding organic matter and aiding in the cycling of nutrients	Requires energy to operate and maintain but has the potential to use less water than soil-based systems

which are soil-based biofacades, grow plants within prefabricated units, whereas hydroponic systems grow plants in a water-based, nutrient-rich medium housed within a container. Wall systems use climbing plants directly clinging to the facades or integrate porous trellises for the plants to climb on.

Modular systems are prefabricated units attached to a building façade and the biofacade modules incorporate frame, growing medium, and vegetation. Panel systems use panels attached to a building facade to support vegetation, made from materials like metal, plastic, or wood. Hydroponic systems grow plants without soil, using a water-based nutrient solution, adaptable for vertical farming and biofacades. Vertical trellis train plants to grow vertically along a support structure, suitable for climbing plants and vines. Planter systems grow plants in containers or planters attached to the facade at various levels to create a green wall effect. Living wall systems grow plants in a specialized medium attached to the facade, which can vary in complexity from simple to intricate. Table 2.3 provides a comparison of the different biofacade systems, their features, benefits, and practical examples.

2.2 Tall Building Enclosure Case Studies in Hot Climates

Climate changes and anthropogenic pollution are increasing due to global population growth and urbanization. This transformation affects health, ecosystems, and the built environment. With limited urban land to accommodate a growing population, many cities worldwide have adopted skyscrapers as their primary form of development. However, these buildings tend to consume significant energy and release carbon dioxide, relying heavily on machines for cooling, ventilation, and lighting. Consequently, efforts have been made to create green tall structures based on sustainable design principles. These strategies employ nature-based solutions and regenerative technologies to reduce environmental impacts.

New building systems and materials are emerging to optimize tall building sustainability performance. High-strength, low-carbon concrete is made of using emerging materials such as CarbonCure and Geopolymer Concrete. CarbonCure technology injects recycled carbon dioxide into concrete mixtures, which helps strengthen concrete and reduce carbon emissions [117]. Geopolymer concrete uses industrial by-products such as fly ash instead of traditional Portland cement, resulting in a significant drop in carbon dioxide emissions [118].

Mass timber and cross-laminated timber (CLT) are nature-based structural materials that perform well structurally and have a low environmental impact. Glulam is an example of mass timber construction where pieces of dimensional lumber are bonded together using high-strength, water-resistant adhesive. It is designed for flexibility and structural strength in tall building systems. Cross-Laminated Timber (CLT) panels are large-sized units made from several layers of timber bonded in alternating grain directions using adhesives, offering durability, fire resistance, and strength. These materials store carbon and have reduced embodied carbon compared to other structural systems such as concrete and steel.

Furthermore, innovations include rainwater harvesting via modern techniques in greywater recycling systems and irrigation. Such measures not only reduce water consumption but also help urban areas become more environmentally friendly

Table 2.3 Comparison of different biofacade systems depending on construction techniques [115, 116]

System type		Description	Performance characteristics/ differentiators
Modular systems	Pod systems	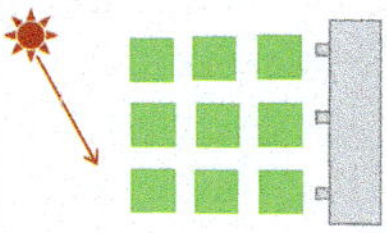	Green wall pods are installed as complete units with a selection of plants; they offer flexibility in design and plant selection, ease of installation, and simplified maintenance
	Planter systems	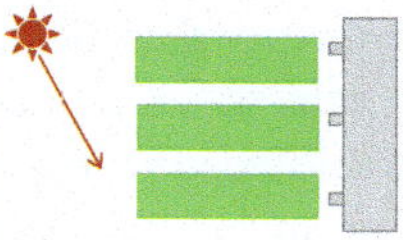	Green wall planters are attached to building facades at different heights with various plant species, enabling easy maintenance and plant replacement; they are used to create patterns and designs, providing additional insulation
	Panel systems	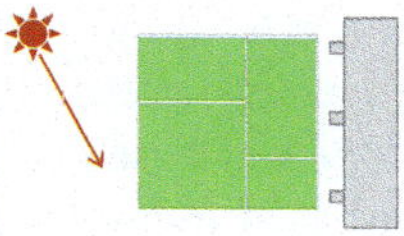	Green wall panels use backup panels integrated with irrigation systems; they provide sturdy support for plants, customizable in size and shape, easy to install
Wall systems	Vertical trellis systems	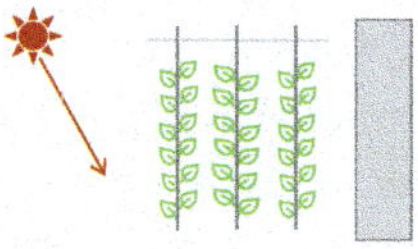	Trellis green walls use trellises made from metal or wood with plants like ivy or climbing roses. They provide a natural aesthetic, shade, privacy, and support for various plant species
	Living wall systems	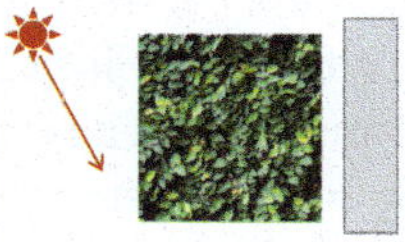	Living wall systems can adopt a variety of plant species; they provide visually appealing facade, improve air quality, insulation, and noise reduction
	Hydroponic systems	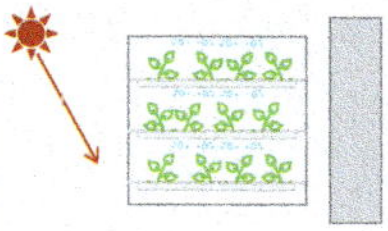	Vertical hydroponic panels with integrated irrigation systems offer efficient water usage, precise control over plant nutrition, and can be easily integrated into various designs

overall. Water management strategies are evolving to include more efficient irrigation systems, such as rainwater harvesting and greywater treatment plants that use efficient irrigation schemes. These closed-loop water systems support landscaping and promote biodiversity within dense urban areas.

Renewable energy integration is crucial for providing clean and sustainable power supplies for building operations. Building-Integrated photovoltaics (BIPV) are an example of advancements in net zero energy architecture, which incorporates solar panels with building enclosures. BIPV also aims to enhance the aesthetic quality of buildings for better public experiences while maximizing energy generation from available vertical surfaces. Facade-integrated PV uses solar panels incorporated directly into the building façades to generate electricity and act as high-performance cladding simultaneously. Roof-integrated PV places solar panels on rooftops, reducing reliance on grid power. BIPV provides cleaner, onsite generated power through building envelopes.

Green walls effectively improve air quality near buildings. These living facades filter pollutants, minimizing negative effects on human health and the environment, thereby creating healthier cities. Biofacades typically incorporate vegetation on planter structures, building walls, or specially designed supporting structures. They are systems in which building enclosures are transformed into living walls through modular planting systems or trellis systems hung from building facades. Sky gardens consist of elevated green spaces filled with vegetation, usually integrated into skyscraper design; these can include green terraces, green roofs, or open spaces between floors. Living walls can be soil-based or hydroponic, and they cover building surfaces with vegetation. Soil-based living walls provide natural growing medium for plants and essential nutrients. They are heavier than hydroponic systems and require stronger structural support. Hydroponic living walls incorporate nutrient rich media and are generally lighter than soil-based green walls, making them suitable for a wider range of structural applications. Their operation and maintenance rely on advanced technologies, such as automated irrigation controllers, nutrient dispensers, environmental sensors, and remote monitoring systems.

The following section analyzes tall buildings located in hot climate regions, focusing on sustainable facade design principles. These include bioclimatic design, integration of energy-efficient building service systems, renewable energy integration, and nature integration, all aimed towards regenerative, carbon-neutral built environments. Figure 2.2 illustrates the locations of case study buildings on a world map.

2.2.1 Tall Building Enclosures in Hot Climates: Asia

(1) **Kanchanjunga Apartments, Mumbai, India, 1983**

Mumbai, India, is a major financial center with great historical significance [119]. It is closely associated with Mahatma Gandhi's political activities and served as an active port during the British colonial period. With a population of over 12 million as of 2024, Mumbai faces urbanization challenges that include a large population of slum dwellers and limited resources [120]. Mumbai has a tropical savanna climate (Köppen: Aw), characterized by distinct wet and dry seasons, with an average annual

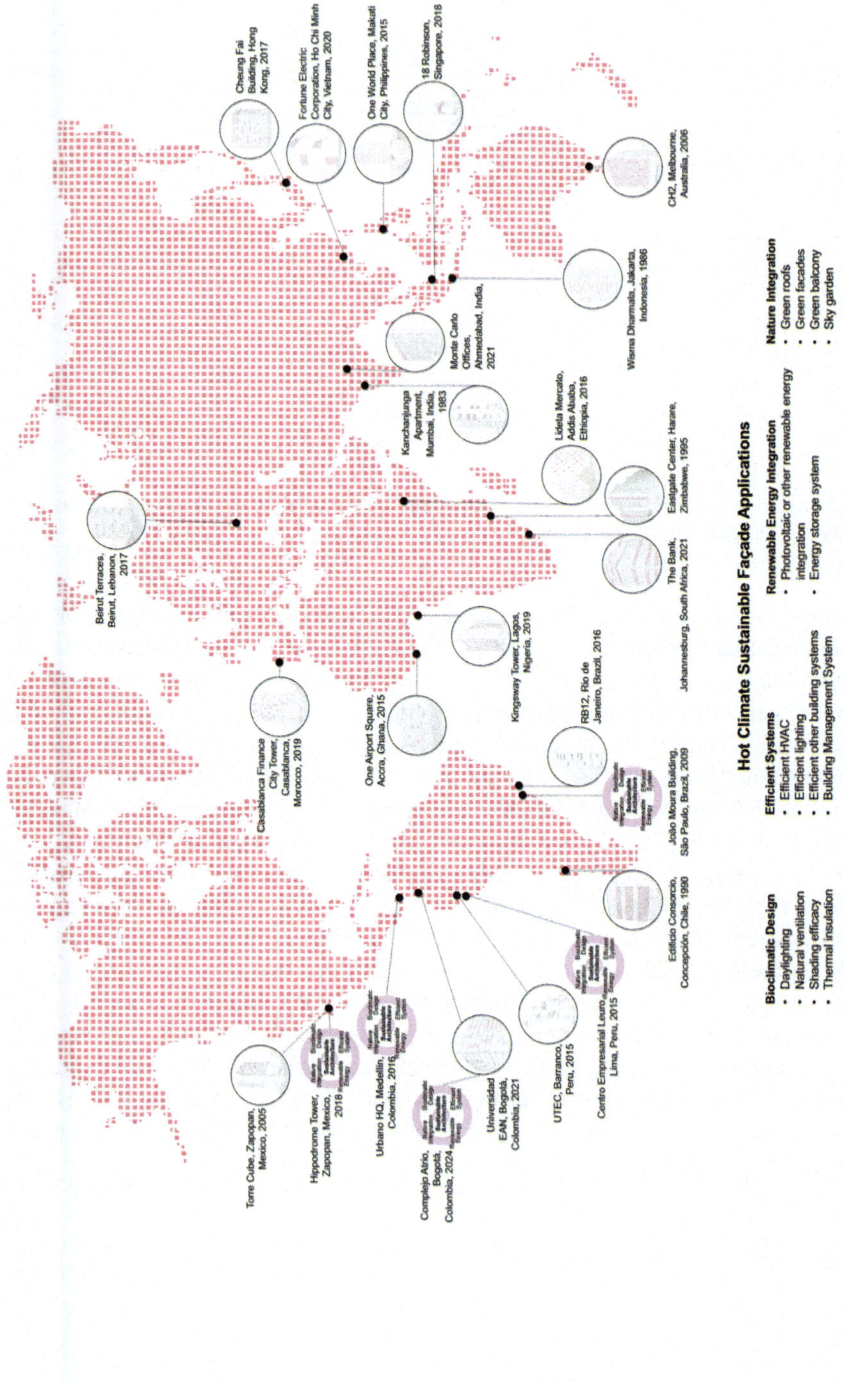

Fig. 2.2 Global case studies of hot climate tall buildings and sustainable façade applications

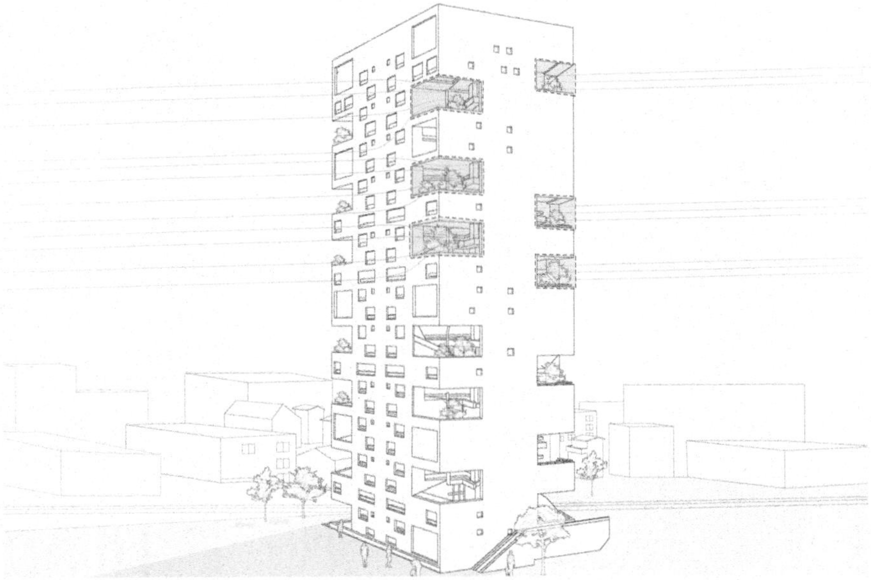

Fig. 2.3 Exterior view of Kanchanjunga Apartments in Mumbai, India (drawings reproduced courtesy of Charles Correa Foundation)

temperature of 26.8 °C (80.2 °F) and the annual rainfall of 2386 mm (93.9 in) [121]. Mumbai has four distinct seasons: (1) Monsoon (June–September): intense rainfall and high humidity, (2) Post-monsoon (October–November): reduced rain accompanied by generally pleasant weather, (3) Winter (December–February): cool and dry conditions, and (4) Pre-monsoon (March–May): increasing temperatures with occasional thunderstorms. These climatic conditions significantly impact infrastructure, urban development, and peoples' lifestyles in the city, requiring substantial adaptations to the heavy monsoon rains and high levels of humidity (Fig. 2.3).

The Kanchanjunga Apartments in Mumbai were designed by renowned architect Charles Correa in 1973, and the building is considered a masterpiece of contemporary Indian architecture that defines Mumbai's skyline [122]. The building uses locally sourced materials such as red brick walls with carved stone detailing to give Kanchanjunga its distinctive look. The apartments incorporate two courtyards that provide natural light and cross ventilation to the living units without compromising privacy. This design integrates Indian residential units along with the modern amenities, allowing for comfortable living environments. Kanchanjunga emphasizes the balance between luxury and local cultures in sustainable urban architecture (Figs. 2.4 and 2.5).

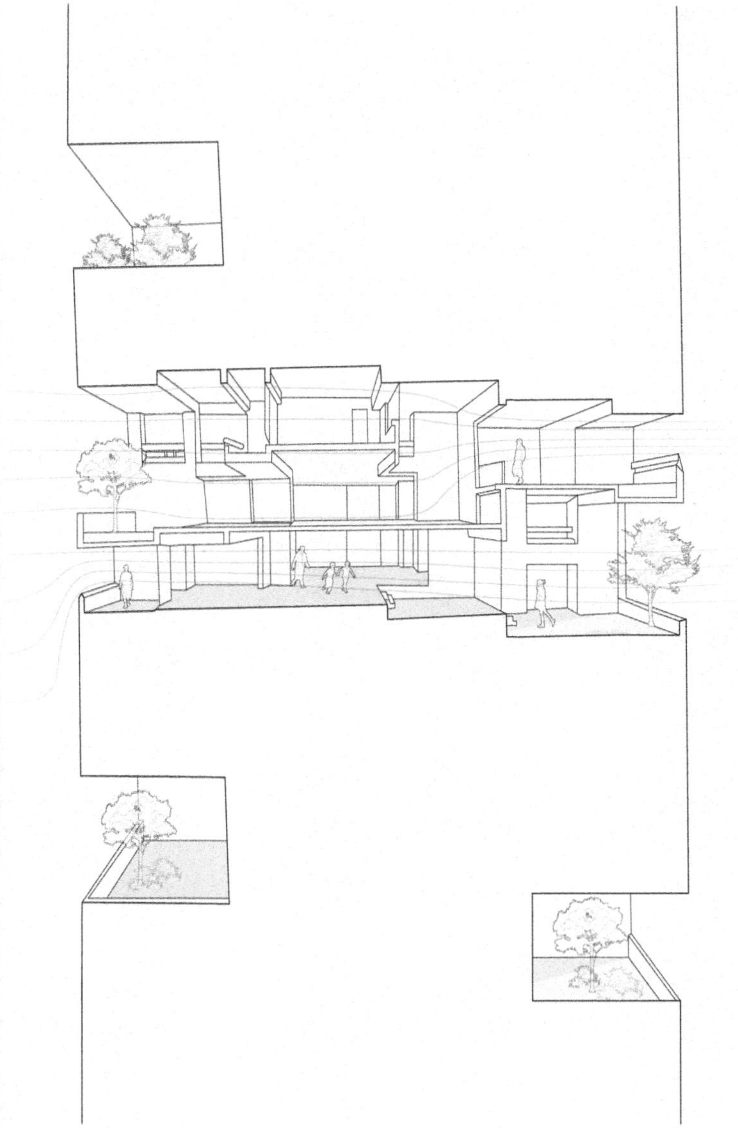

Fig. 2.4 Building section view of Kanchanjunga Apartments in Mumbai, India (drawings reproduced courtesy of Charles Correa Foundation)

Fig. 2.5 Kanchanjunga Apartments, Mumbai, India (1983), and sustainable façade applications

Location	Mumbai, India
Köppen Climate Classification	Tropical Savanna Climate (Köppen: Aw)
Completion Date	1983
Building Owner/Primary Tenant	Kanchanjunga Cooperative Housing Society
Architects & Design Teams	Charles Correa
Number of Floors	32
Gross Floor Areas	6,500 m^2

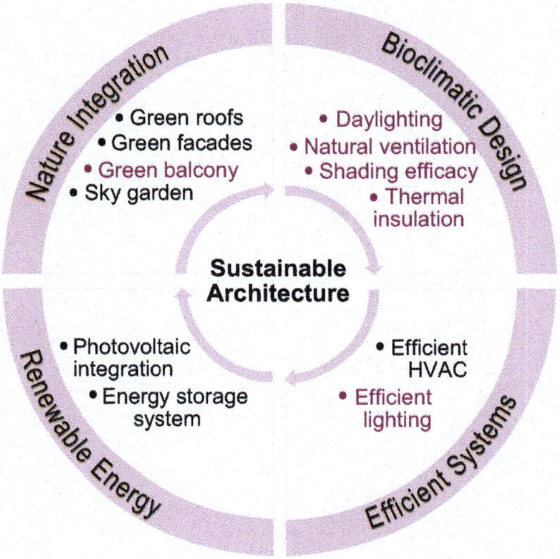

(2) **Wisma Dharmala, Jakarta, Indonesia, 1986**

Jakarta, a metropolis in Southeast Asia and the capital city of the Republic of Indonesia had a population of 11.4 million people in 2024, and covers a city land area of 4384 km^2 (1693 mi^2) [123]. The history of the city began as a fishing region in the fourth century which later developed into a trading hub that drew merchants from all over the world; the city later subsequently came under the control of various empires before becoming part of Indonesia in 1945, with the indigenous Betawi people developing a rich culture after centuries of intermarriage [124]. The city can be defined as a tropical monsoon climate (Köppen: Am), characterized by short dry season and intense monsoon rains, with high temperatures and humidity year-round [125]. The average temperature ranges approximately 27.6 °C (81.7 °F), with yearly precipitation of about 1855 mm (73 in). November to March is characterized as the wet season, during which the area receives heavy rainfall, whereas June–September is the dry season in which relatively little rain falls but the weather remains hot and humid. May–June and October–November serve as transitional months between dry and wet periods. This climate results in high levels of moisture and rainfall, which leads to rich plant growth and influences the urban planning and architectural style of buildings in the city (Figs. 2.6 and 2.7).

Wisma Dharmala is an office building located in Jakarta designed by the American architect Paul Rudolf, who was one of the most notable figures in modern architecture in the latter half of the twentieth century. Wisma Dharmala is also distinguished by its utilitarian design, made up of simple lines and geometric shapes, which differentiate it from the adjacent high-rise buildings [126]. The facade of the building incorporated local Stone and large windows for daylighting and views over the city [127]. Wisma Dharmala, the geometric architectural form and functional space layout, has become an inspiration to Indonesian architects. Many architects have incorporated similar design approaches, such as a practical architectural solution, the use of locally available materials, and the provision of daylight. The building continues to be relevant, inspiring architects and remaining a successful high-rise development [128] (Fig. 2.8).

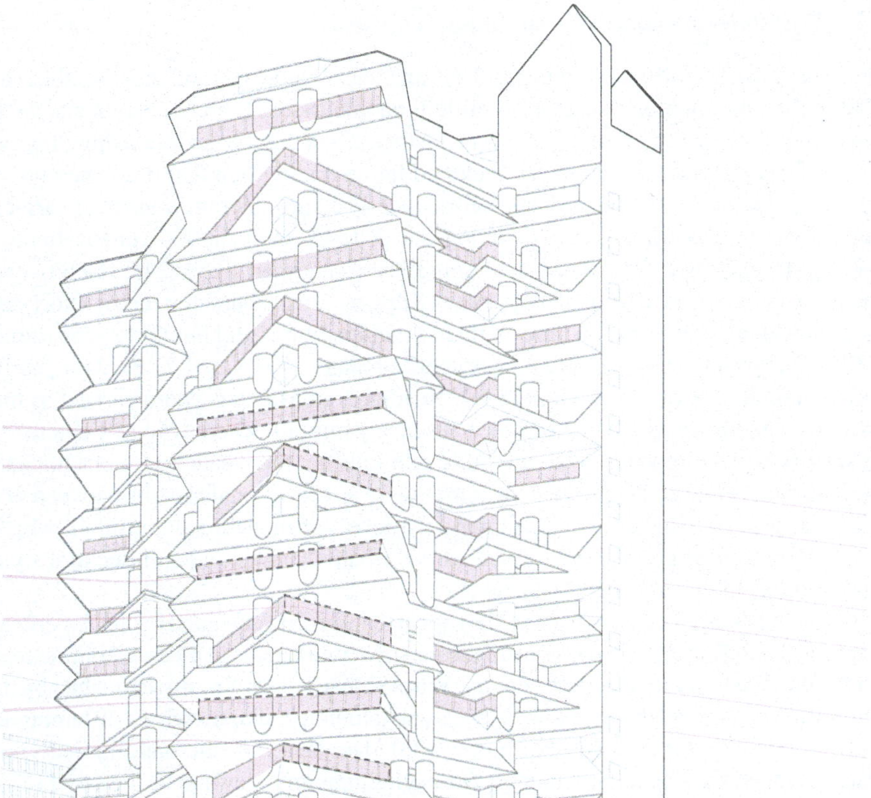

Fig. 2.6 Exterior view of the Wisma Dharmala in Jakarta, Indonesia (drawings reproduced courtesy of Paul Rudolph Foundation)

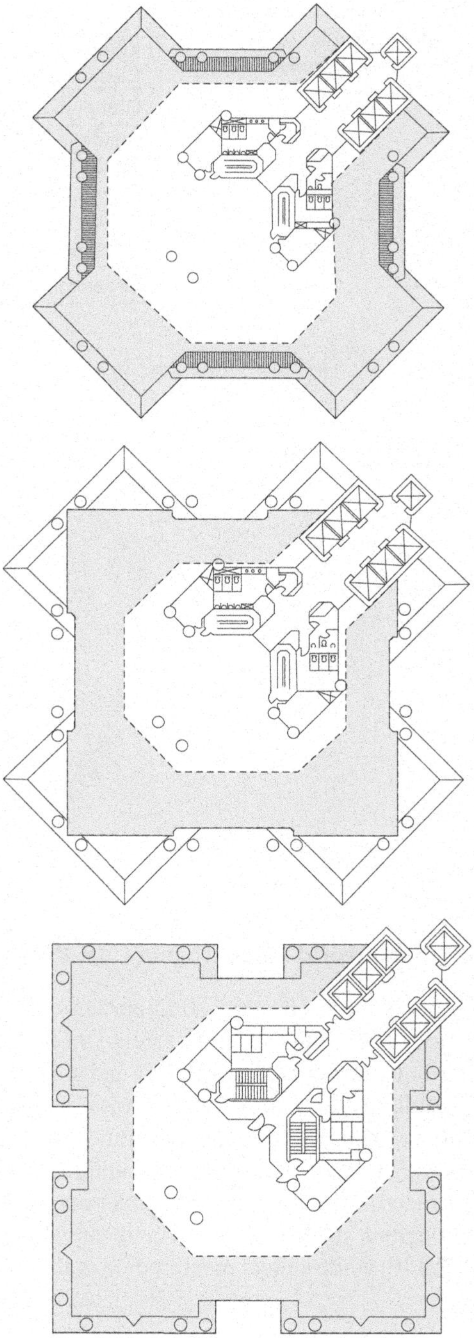

Fig. 2.7 Floor plans of Wisma Dharmala in Jakarta, Indonesia (drawings reproduced courtesy of Paul Rudolph Foundation)

Fig. 2.8 Wisma Dharmala, Jakarta, Indonesia (1986), and sustainable façade applications

Location	Jakarta, Indonesia
Köppen Climate Classification	Tropical Monsoon Climate (Köppen: Am)
Completion Date	1988
Building Owner/Primary Tenant	Wisma Dharmala Sakti Office Headquarters
Architects & Design Teams	Paul Rudolph
Number of Floors	23
Gross Floor Areas	30,986 m^2

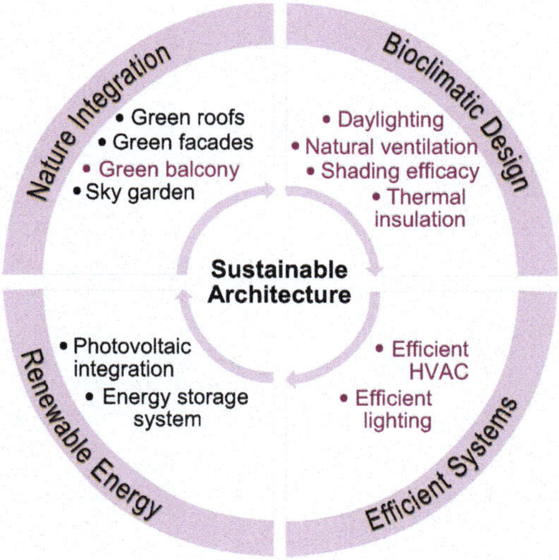

(3) **One World Place, Taguig City, Philippines, 2015**

Taguig City has become an important emerging business district in Southeast Asia largely due to the support of foreign investors, master-planned urban development in in Bonifacio Global City (BGC), and non-tariff measures such as tax policy [129]. As a result, several multinational firms have established their regional offices in Taguig City, making the area a major hub for commercial and landmark activities within Metro Manila. Taguig City lies within a tropical savannah climate (Köppen: Aw), characterized by monsoon summer and dry winter with warm to hot temperatures throughout the year [130]. The mean yearly temperature is approximately 26.6 °C (79.9 °F) while annual precipitation is about 1666 mm (65.6 in). The months between

June and November are rainy months, while the remaining months from December to May are relatively dry with occasional rainfall, making the months humid (Figs. 2.9 and 2.10).

Fig. 2.9 Exterior view of One World Place in Taguig City, Philippines (drawings reproduced courtesy of Gensler)

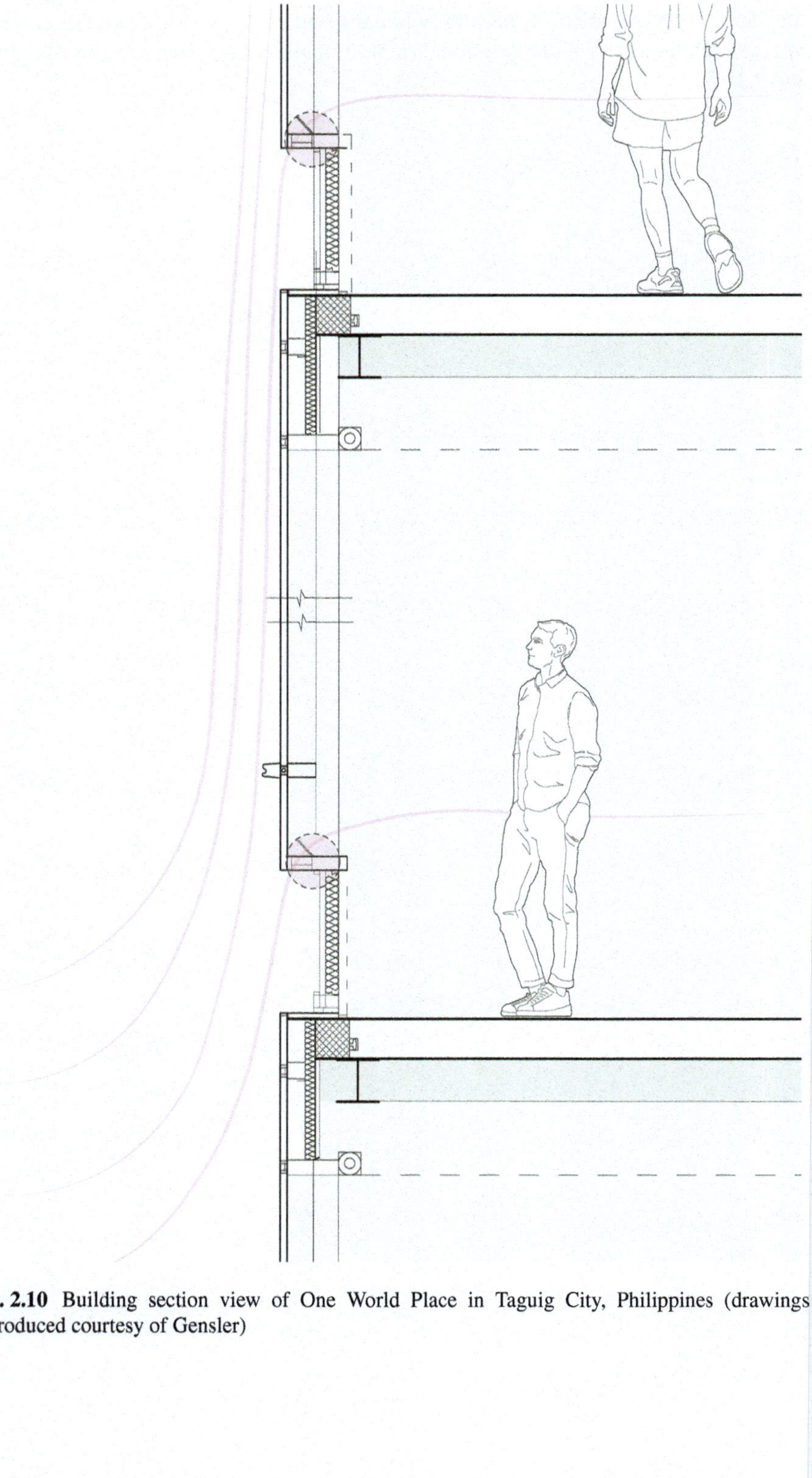

Fig. 2.10 Building section view of One World Place in Taguig City, Philippines (drawings reproduced courtesy of Gensler)

One World Place is a Class A office building located in Taguig City, one of the busiest business districts within Metro Manila in the Philippines [131]. Its strategic site location makes it attractive to businesses as a prime real estate, ensuring easy access from all parts of Metro Manila. The interior features help create an interactive workplace environment for the office occupants while providing comfortable spaces that accommodate individual needs. The interiors consist of a modern architectural design with high ceilings, spacious lobbies, and fast elevators. The high-performance building envelope integrates micro-louvres within prefabricated curtain wall systems for natural ventilation and passive cooling. The facades allow light to be transmitted and penetrate deep into interior space. Daylight and naturally ventilated facades improve the indoor environment quality (IEQ), which in turn enhances the productivity and health of the users. Several retail units within the complex offer occupants and visitors convenient access to breaks or lunch during their working hours (Fig. 2.11).

Fig. 2.11 One World Place, Taguig City, Philippines (2015), and sustainable façade applications

Location	Taguig City, Philippines
Köppen Climate Classification	Tropical Savanna Climate (Köppen: Aw)
Completion Date	2015
Building Owner/Primary Tenant	Daiichi Properties
Architects & Design Teams	Gensler; Aidea (local architect)
Number of Floors	31
Gross Floor Areas	49,374 m^2

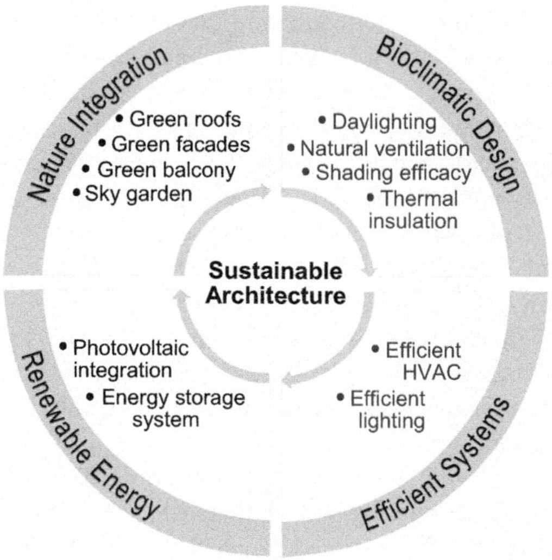

Makati City has become an important financial district in Southeast Asia mainly due to the support of foreign investors, land reclamation projects, and non-tariff measures like tax policy [63]. As a result, several multinational firms have also set up their regional offices in Makati making it a regional hub for commercial and iconic activities. Makati City lies within the tropics and savannah climate type (Köppen: Aw) characterized by high humidity and hot temperature throughout the season [64]. The mean yearly temperature has been reported to be 26.6 °C while annual precipitation stands at 1666 mm. The months between June and November are rainy months with the rest of the months from December to May dry months with little but some rain, making the months humid.

One World Place is a Class A office building located in Makati City, one of the busiest business districts within Metro Manila in the Philippines [65]. Its strategic site location makes it desirable to businesses as prime real estate location ensuring easy access from all parts of Metro Manila. The office features help create an interactive workplace environment for the office occupants while providing comfortable space for their individual needs. The interiors consist of modern architectural design with high ceilings, spacious lobbies and fast elevators. High performance building envelope integrates micro-louvres within prefabricated curtain wall systems for natural ventilation and passive cooling. The facades allow light to be transmitted and bring deep into the space. Daylight and naturally ventilated facades improve the indoor environment quality (IEQ), which in turn stimulates the productivity and health of the users. A number of the retail units within the compounds offer the occupants and the visitors the ease of taking breaks or lunch during their working hours.

(4) **Beirut Terraces, Beirut, Lebanon, 2017**

Beirut, the capital of Lebanon, has 2.4 million residents as of 2024, and the metropolitan area covers 67 km^2 (25.9 mi^2) [132]. Beirut has evolved into Lebanon's financial center while maintaining its cultural heritage, and following French rule during World War I, it became the capital of an independent Lebanon in 1943 [133]. Despite the damages caused by civil wars, Beirut has shown economic growth since 2005, attracting foreign investment and creating jobs [134]. Beirut has a hot-summer Mediterranean climate (Köppen: Csa), characterized by hot, dry summers and mild, wet winters [135]. Summer temperatures average around 26.1 °C (79 °F), while winter temperatures average about 10.5 °C (50.9 °F). Most rainfall occurs from November to March, with summers being dry. The annual precipitation averages 726 mm (28.6 in) (Figs. 2.12 and 2.13).

The Beirut Terraces concept combines respect for local context with modern design, based on five design principles: layered terraces, indoor-outdoor living, greenery, private views, and a light-driven identity [136]. According to the architect, the building offers diverse apartment types, including simplex, duplex, and townhouse units with private and amenity pools. The design features varied slab sizes, creating flexible living spaces that balance openness and privacy. Environmental engineering and integrated vegetation enhance sustainability. Large overhangs and strategic perforations reduce solar gains and control light. Thick floor plates provide thermal mass, helping to balance daily temperature cycles. This modern

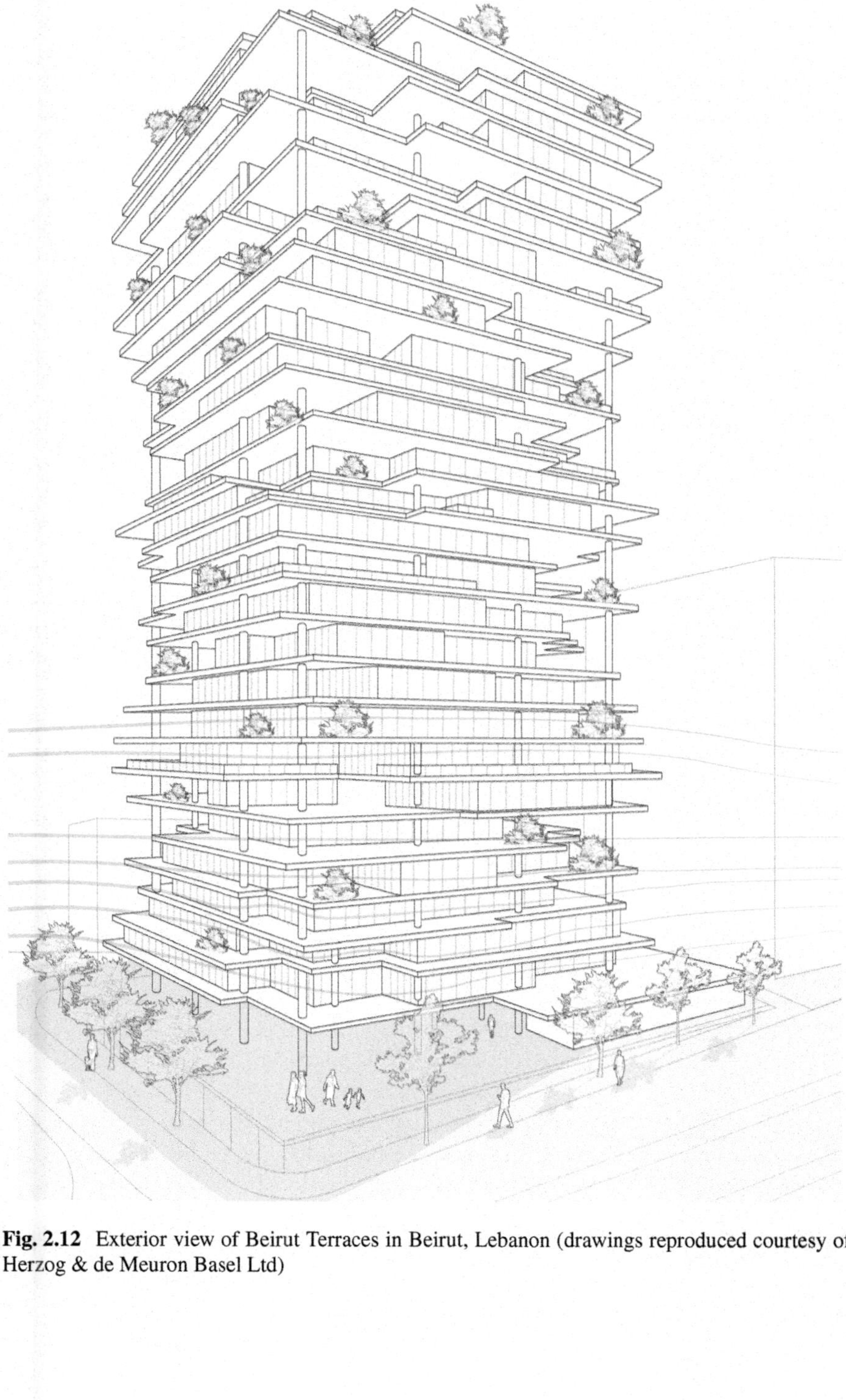

Fig. 2.12 Exterior view of Beirut Terraces in Beirut, Lebanon (drawings reproduced courtesy of Herzog & de Meuron Basel Ltd)

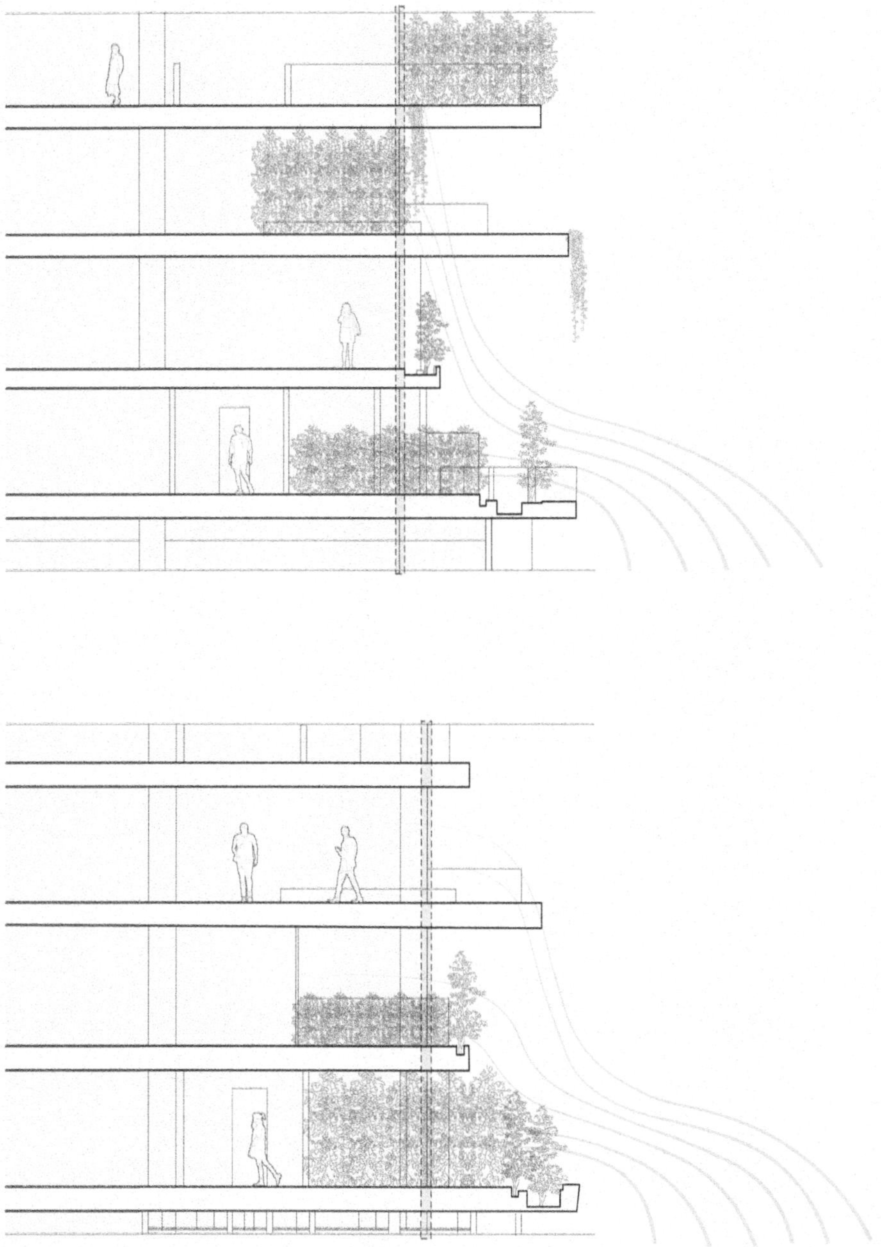

Fig. 2.13 Building section view of Beirut Terraces in Beirut, Lebanon (drawings reproduced courtesy of Herzog & de Meuron Basel Ltd)

Fig. 2.14 Beirut Terraces, Beirut, Lebanon (2017), and sustainable façade applications

Location	Beirut, Lebanon
Köppen Climate Classification	Hot-summer Mediterranean Climate (Köppen: Csa)
Completion Date	2017
Building Owner/Primary Tenant	Madaïn Properties
Architects & Design Teams	Herzog & de Meuron
Number of Floors	26
Gross Floor Areas	79,200 m^2

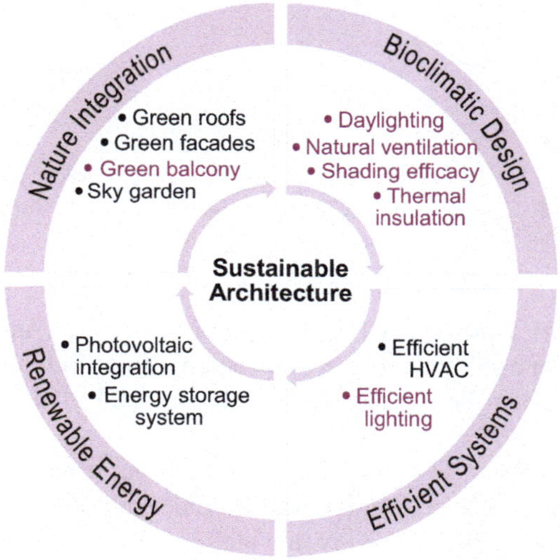

green building technology creates balcony gardens on high-rises, reducing energy consumption, improving air quality, providing wildlife habitats, and enhancing mental health through contact with nature [137] (Fig. 2.14).

(5) Cheung Fai Building, Hong Kong, 2017

Hong Kong was a British colony until 1997 and has become a metropolitan city famous for blending eastern and western cultures as well as a global financial hub, due to low taxes and strong legal structures [138]. The population of the metropolis city covering 1104 km^2 (426.3 mi^2) and had 7.4 million people as of 2024, and is densely populated and multilingual with Cantonese and English as the two official languages [139]. Hong Kong comprises mainly urban regions alongside urban parks and natural landscapes. Hong Kong has a humid subtropical climate (Köppen: Cfa), characterized by hot humid summers and mild winters with an average temperature of 22.6 °C (72.7 °F) and the annual rainfall of approximately 1796 mm (70.7 in) [140].

Fig. 2.15 Exterior view of Cheung Fai building in Hong Kong (drawings reproduced courtesy of MVRDV)

The city experiences four distinct seasons; these include a hot summer spanning from June to September, a warm autumn from October up to early December, a mild winter from late December to February, and a humid spring from March to May (Fig. 2.15).

The Cheung Fai Building in Hong Kong is an adaptive reuse project that transforms warehouse to an innovative office environment [141]. Designed by the Dutch architecture firm MVRDV, the building blends traditional industrial elements with contemporary design. The new architectural features utilize various materials such as glass and steel while maintaining harmony with nature through its incorporation of green spaces. The building showcases innovative design aspects, including the use of daylight, natural ventilation, solar panels on the roof, and special ventilation systems that help cool each room. Its creative design and lively ambiance have transformed it into a notable tourist attraction, distinguishing it from neighboring buildings. These factors contribute to making the Cheung Fai Building one of the most sustainable adaptive projects in Hong Kong (Fig. 2.16).

Fig. 2.16 Cheung Fai building, Hong Kong (2017), and sustainable façade features

Location	Hong Kong, China
Köppen Climate Classification	Humid Subtropical Climate (Köppen: Cfa)
Completion Date	2015
Building Owner/Primary Tenant	GAW Capital Partners
Architects & Design Teams	MVRDV
Number of Floors	14
Gross Floor Areas	18,000 m² (façade surface area for the project)

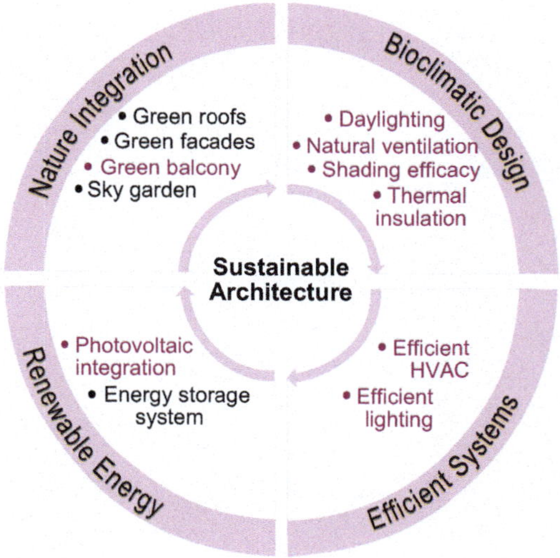

(6) **18 Robinson, Singapore, 2018**

Singapore has a population of approximately 6.1 million people in 2024, living on just 734 km^2 (283.4 mi^2) of land that includes one main island surrounded by over 60 smaller ones at the southern tip of the Malay Peninsula [142]. Singapore's policies have enabled rapid industrialization and foreign investment, resulting in efficient utilization of resources and job creation through programs such as public housing [143]. This has led to low unemployment rates, while GDP per capita has increased from approximately USD 500 in 1965 to nearly USD 50,000 today (Figs. 2.17 and 2.18).

The climate of Singapore is tropical rainforest climate (Köppen: Af), characterized by no dry season and heavy rainfall year-round, with an average annual temperature of 26.7 °C (80.1 °F) and humidity levels exceeding over 80% year-round, accompanied by heavy rainfall of approximately 2366 mm (93.1 in) per year and no distinct dry season [144]. Located near the equator, it has a small temperature range and plenty

Fig. 2.17 Exterior view of 18 Robinson in Singapore (drawings reproduced courtesy of Kohn Pedersen Fox)

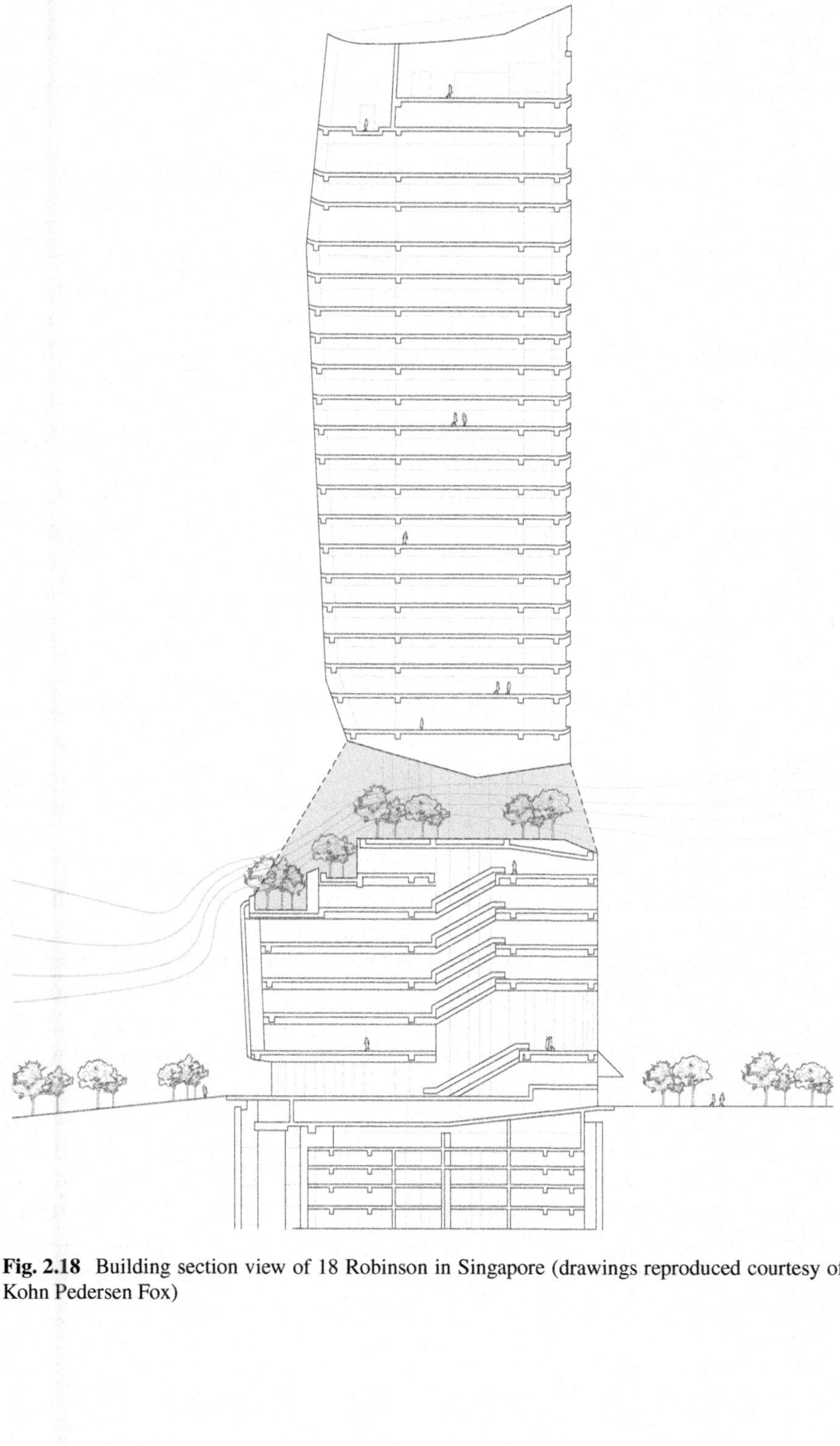

Fig. 2.18 Building section view of 18 Robinson in Singapore (drawings reproduced courtesy of Kohn Pedersen Fox)

Fig. 2.19 A 18 Robinson, Singapore (2018), and sustainable façade applications

Location	Singapore
Köppen Climate Classification	Tropical Rainforest Climate (Köppen: Af)
Completion Date	2018
Building Owner/Primary Tenant	Tuan Sing Holdings Limited
Architects & Design Teams	Kohn Pedersen Fox Associates (KPF)
Number of Floors	28
Gross Floor Areas	12,170 m^2

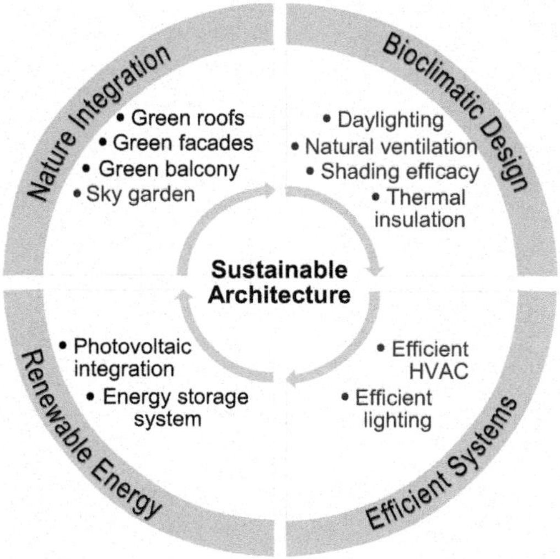

of sunshine interspersed with frequent thunderstorms. Singapore's tropical climate promotes biodiversity, giving rise to its reputation as a "Garden City."

The 18 Robinson building, designed by Kohn Pedersen Fox Associates, is located in Singapore's Central Business District [145]. Large urban windows on Robinson Road showcase internal activity, while the podium's design complements Lau Pa Sat's angular roof and terracotta features. Its glass exterior provides panoramic views and reflects the cityscape. 18 Robinson responds to its unique site shape through a balanced, faceted tower design. The building comprises a crystalline office tower and a retail podium. It offers practical benefits, including advanced seismic technology for disaster resistance, energy-efficient lighting, and automated climate control systems. These features ensure comfort, safety, and lower utility costs for tenants. Following Singapore's Landscape Replacement policy, it provides public green spaces equal to the site area, including a landscaped podium and rooftop sky garden with trees, natural ventilation, and city views [146] (Fig. 2.19).

(7) Fortune Electric Corporation, Ho Chi Minh City, Vietnam, 2020

Ho Chi Minh City, previously known as Saigon, was renamed in 1975. It had approximately three million inhabitants in 1995 and has grown to approximately 9.5 million as of 2024 within the territory of 2061 km^2 (795.8 mi^2) [147]. This remarkable expansion has also led to the development of infrastructure supporting businesses and transportation networks. The growth of the tourism business created further opportunities for hotels and restaurants, particularly in the proximity of the airports. Such high density, rapid urbanization has caused congestion issues. Several municipal authorities have slowed further rapid urbanization to find appropriate solutions to these problems and implement effective city planning and urban development (Figs. 2.20 and 2.21).

Ho Chi Minh City is tropical climates with an average daily temperature of approximately 27 °C (80.6 °F) throughout the year [148]. It falls under a tropical savanna climate (Köppen: Aw) with distinct rainy and dry seasons. Thus, heavy rains and high humidity are experienced from May to November, while lighter rainfall occurs from December to April. This climate supports the growth of trees and vegetation, which is evident throughout the city.

Located in Ho Chi Minh City, Vietnam, the Nanoco Head Office is a notable building that highlights contemporary architecture [149]. Designed by VTN Architects designed, the glass block tower houses Fortune Electric's headquarters and showroom. VTN Architects, known for their sustainable design solutions, implemented a glass block exterior, which makes the tower an iconic element in the city. The interplay of the building's form and exterior articulation creates an impressive visual experience, both complementing and contrasting the surrounding cityscape. The glass block façade and green balcony filter sunlight, creating a soft, ambient light while minimizing solar heat gain. Additionally, these sustainable elements enhance acoustic performance by reducing city noise entering the space (Fig. 2.22).

Fig. 2.20 Exterior view of Fortune Electric Corporation in Ho Chi Minh City, Vietnam (drawings reproduced courtesy of VTN Architects)

Fig. 2.21 Building section view of Fortune Electric Corporation in Ho Chi Minh City, Vietnam (drawings reproduced courtesy of VTN Architects)

Fig. 2.22 Fortune Electric Corporation in Ho Chi Minh City, Vietnam (2018), and sustainable façade applications

Location	Ho Chi Minh City, Vietnam
Köppen Climate Classification	Tropical Savanna Climate (Köppen: Aw)
Completion Date	2018
Building Owner/Primary Tenant	Fortune Electric Corporation
Architects & Design Teams	VTN Architects
Number of Floors	10
Gross Floor Areas	2,191 m^2

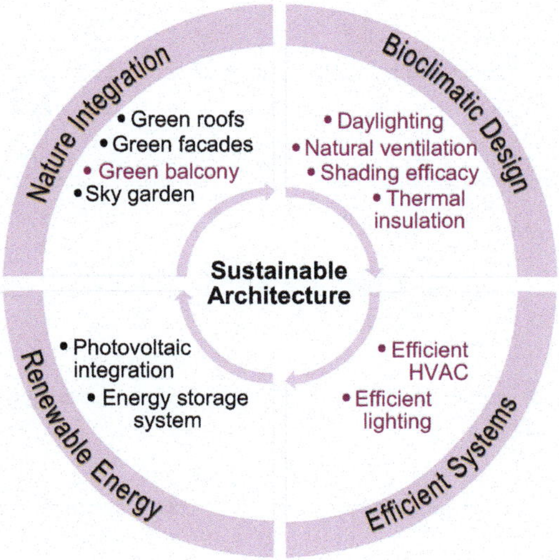

(8) **Monte Carlo Offices, Ahmedabad, India, 2021**

Ahmedabad is home to approximately 8.8 million people as of 2024, spread across about 464 km^2 (179.2 mi^2) [150]. Established in 1411 by Sultan Ahmad Shah as a prominent city in Gujarat in India, it was well known for its textile industry during British rule [151]. In the urban development studies, four major eras can be identified in terms of its growth: pre-independence, post-independence, economic liberalization, and recent times. In each phase, there were marked transformations, such as the construction of new infrastructure and rapid population increase due to industrialization, among other factors. At present, Ahmedabad is amongst India's most developed cities, with investments coming from abroad mainly into the information technology sector. Ahmedabad has a hot semi-arid climate (Köppen climate: BSh), characterized by low rainfall and hot temperatures [152]. Summers (March–June) are hot with high average temperatures, with a recorded maximum of 33 °C (91.4 °F).

Fig. 2.23 Exterior view of Monte Carlo offices in Ahmedabad, India (drawings reproduced from Edifice Consultants)

The monsoon season (July–September) brings heavy rainfall. Winters (November–February) are mild, with an average temperature of approximately 20.4 °C (68.7 °F). Annual precipitation is approximately 757 mm (29.8 in) (Fig. 2.23).

The Monte Carlo Offices in Ahmedabad, designed by Edifice Consultants, showcase modern architecture. The building's construction results in a sleek design that distinguishes it from neighboring structures, reflecting Ahmedabad's historic walled-city heritage [153]. According to the Architect, the city's layout, resembling a patchwork quilt of villages, inspired the façade's quilted pattern and resulted in triangular tessellations resembling the Uttarayan kite festival. The design concept also references Ahmedabad's history as a diamond-cutting center, with a façade mimicking diamond facets to maximize natural light. The glass envelope, composed of rhomboid fractal geometries, creates a modern interpretation of the city's heritage [154]. The purposeful placement of indented facade elements creates interesting visual effects throughout the building's exterior while providing daylighting penetration, solar regulation, and a enhanced user experience (Figs. 2.24 and 2.25).

Fig. 2.24 Building section view of Monte Carlo offices in Ahmedabad, India (drawings reproduced courtesy of Edifice Consultants)

Fig. 2.25 Monte Carlo offices in Ahmedabad, India (2021), and sustainable façade applications

Location	Ahmedabad, Gujarat, India
Köppen Climate Classification	Hot Semi-Arid Climate (Köppen: BSh)
Completion Date	2021
Building Owner/Primary Tenant	Monte Carlo Headquarter
Architects & Design Teams	Edifice Consultants Pvt. Ltd
Number of Floors	12
Gross Floor Areas	5,574 m^2

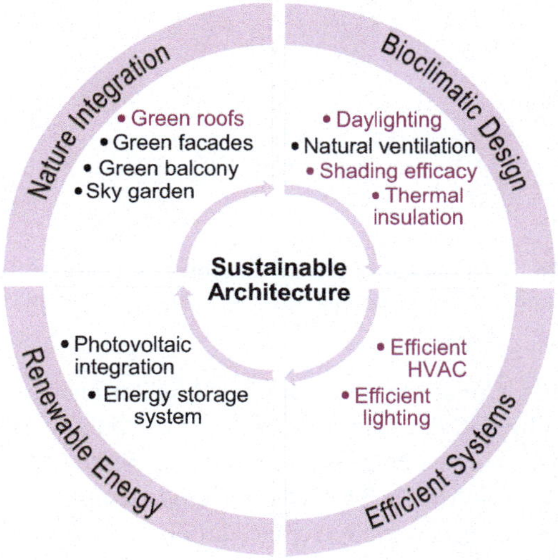

2.2.2 *Tall Building Enclosures in Hot Climates: South America*

(1) **Edificio Consorcio Nacional de Seguros, Santiago, Chile, 1990**

Located in the south-central of Chile, Santiago is the capital and largest city of Chile, with approximately seven million metropolitan population as of 2024 [155]. Its strategic location within the Chilean Central Valley, between the Andes Mountains and Coastal Range has increased its importance over time as the political, economic, and financial center of Chile; More recently, areas close to the downtown historic and business districts have developed rapidly due to the construction of new high-density residential and mixed-use developments nearby [156]. The growth in population has brought an expansion of job opportunities but also a more diverse population and greater cultural diversity, making the city one of the most active regions in South America today [157]. Its residents work hard to conserve their culture, history and traditions while also having aspirations of development and progress in the region.

Fig. 2.26 Exterior view of Edificio Consorcio Nacional de Seguros in Santiago, Chile (drawings reproduced courtesy of Enrique Browne & Associates Architects)

Santiago, Chile, has a hot-summer Mediterranean climate (Köppen: Csa), characterized by hot, dry summers and mild to cool, humid winters [158]. It experiences mild and rainy winters between May and September and hot and dry summers from November to March. The lowest average temperature is in July at 3.7 °C (38.6 °F), and the warmest period averages around 28.7 °C (83.6 °F) in January. The annual average temperature is approximately 15.7 °C (60.3 °F), and the total annual rainfall is about 517 mm (20.4 in). The coastal location helps moderate year-round temperatures (Figs. 2.26 and 2.27).

Edificio Consorcio Nacional de Seguros is located in Santiago, Chile, and blends well with the surrounding cityscape. Completed in 2007 and designed by Enrique Browne, the building incorporates curved walls on each floor. The curved walls feature several green spaces such as green terraces and gardens, and provide occupants with access to nature within the urban environment. The natural form of the façade adds to the architectural volume, making it unique within Santiago's skyline. The building further enhances sustainable design through the use of natural ventilation systems and energy-efficient windows to reduce electricity costs while providing an aesthetically pleasing appearance [159]. Edificio Consorcio Nacional de Seguros is an example of combining modern design with traditional Chilean architecture and create a harmony between old and new styles. Geometric shapes, including circular window openings, add visual interest without compromising structure (Fig. 2.28).

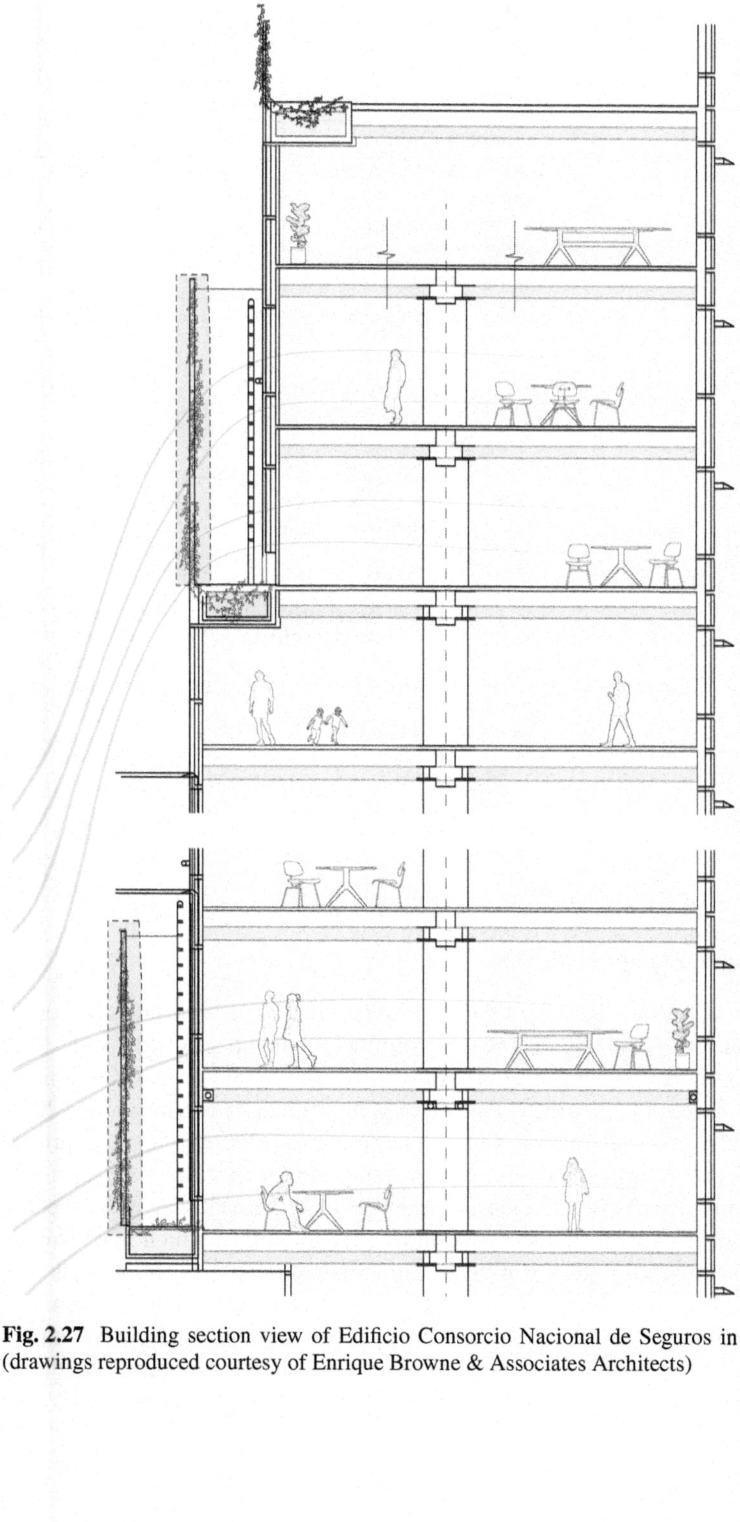

Fig. 2.27 Building section view of Edificio Consorcio Nacional de Seguros in Santiago, Chile (drawings reproduced courtesy of Enrique Browne & Associates Architects)

Fig. 2.28 Edificio Consorcio, Santiago, Chile (1990), and sustainable façade features

Location	Santiago, Chile
Köppen Climate Classification	Hot-Summer Mediterranean Climate (Köppen: Csa)
Completion Date	1993
Building Owner/Primary Tenant	Consorcio Nacional de Seguros
Architects & Design Teams	Borja Huidobro, Enrique Browne
Number of Floors	17
Gross Floor Areas	26,720 m²

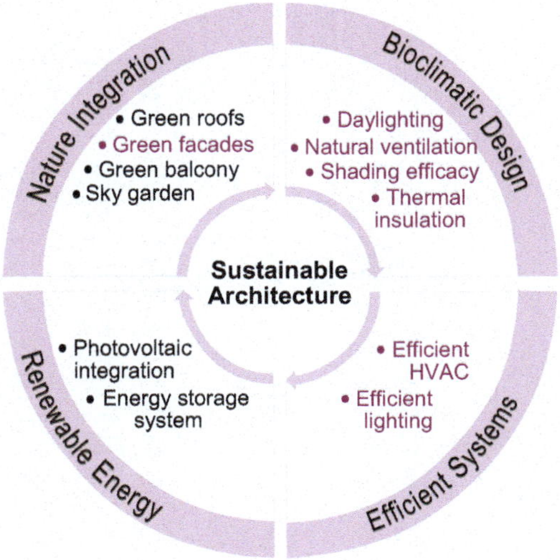

(2) **Torre Cube, Zapopan, Mexico, 2005**

Zapopan, Jalisco, Mexico, has over 1.2 million inhabitants as of 2024 and is part of the metropolitan area of Guadalajara [160]. According to municipal records, it has been developing as a major economic center, mainly in the automobile industry. According to the Climate Action Plan for the Guadalajara Metropolitan Area published in 2020, the city endeavors to achieve carbon neutrality through waste management, improved transportation, and the implementation of more renewable energy [161]. High priority carbon neutrality actions include the introduction of the area's first, fully electric bus route and development of additional bicycle lanes, both of which have already decreased CO_2 emissions and car usage. The climate of Zapopan, Mexico, is hot-summer Mediterranean climate (Köppen: Csa), characterized by hot dry summers and mild wet winters [162]. It is characterized by warm, dry winters from November to April and hot, moist summers from May to October, with most rainfall occurring during summer. The moderate temperature range in the region is influenced by the country's elevation and location, in addition to its annual rainfall of approximately 902 mm (35.5 in) (Figs. 2.29 and 2.30).

La Torre Cube in Zapopan, Mexico, epitomizes contemporary architectural styles. Designed by Luis Bozzo Rotondo, it serves as a landmark of the city by being strategically located at the junction of two important highways [163]. According to the architect's research, La Torre Cube sought to reconstruct an existing dilapidated area as part of a larger urban development. The tower, constructed using concrete and steel, creates a cubical form. Other design features include curved facades and punched windows that provide unobstructed views of the surrounding area. The project promotes sustainability through a high level of recycling materials, such as using beams made from recovered steel in the construction of various building elements. The urban context and the tower's scale presented challenges during the construction process. This combination of classic construction and engineering solutions with sustainable technology enabled fast, high-quality construction (Fig. 2.31).

Fig. 2.29 Exterior view of Torre Cube in Zapopan, Mexico (drawings reproduced courtesy of Luis Bozzo Rotondo)

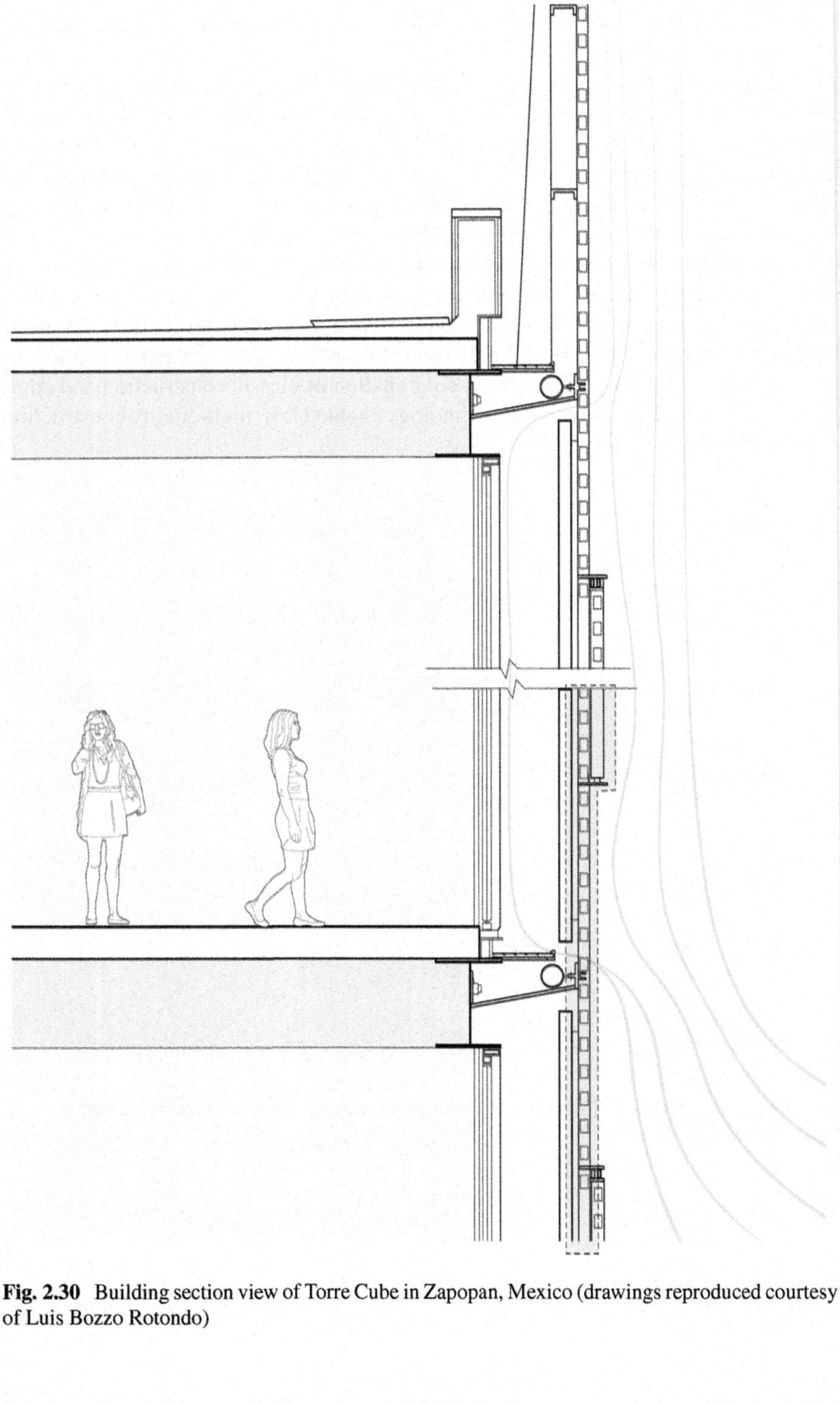

Fig. 2.30 Building section view of Torre Cube in Zapopan, Mexico (drawings reproduced courtesy of Luis Bozzo Rotondo)

Fig. 2.31 Torre Cube, Zapopan, Mexico (2005), and sustainable façade applications

Location	Zapopan, Jalisco, Mexico
Köppen Climate Classification	Hot-summer Mediterranean Climate (Köppen: Csa)
Completion Date	2005
Building Owner/Primary Tenant	Cube Internacional
Architects & Design Teams	Carme Pinos
Number of Floors	16
Gross Floor Areas	17,000 m^2

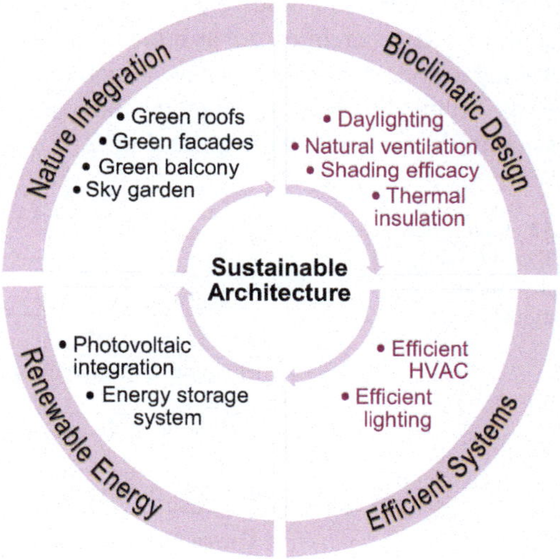

(3) **João Moura Building, São Paulo, Brazil, 2009**

São Paulo, Brazil's largest city had a population of approximately 22.8 million as of 2024 and covers an area of about 1493 km^2 (576.4 mi^2) [164]. The urban sprawl has resulted in complex infrastructure such, as roads, highways, and public transportation systems, and has outpaced coordinated development, leading to traffic congestion and the overburdening of public service systems [165]. Research also indicates that lack of adequate housing and social inequality are also problems in São Paulo, as some parts of the city are extremely affluent while many informal settlements lack basic services for large populations. In recent years, São Paulo has focused on sustainable urban development through the enhancement of public transportation, reduction of pollution levels, and the expansion of green areas [166].

São Paulo, Brazil, has a humid subtropical climate (Köppen: Cfa), characterized by hot humid summers, mild winters [167]. São Paulo's climate features average high temperatures of 22.5 °C (72.5 °F), and low temperatures of 16.2 °C (61.2 °F). Annual rainfall averages 1356 mm (53.4 in), concentrated between October and

March. Humidity is high, especially in summer, with moderate seasonal temperature variations.

The João Moura building is a Class A office building located at the intersection of João Moura Street and Sumaré Avenue, connecting Vila Madalena and Pinheiros districts [168]. According to the architect, the project responds the topography of the land to house subtly while minimizing any earthworks or excavations even in underground spaces. On João Moura Street, the entrance setback is 10 m (32.8 ft), twice the legally required distance, creating a generous forecourt and emphasizing the user experience of entering the building. The north and south facades consist of operable windows and colorful panels to encourage natural ventilation, daylight penetration, and solar regulations. The terraced garden provides green access, air quality improvement, and biodiversity. The project uses topography and strategic site location to assert architecture as a mediator between the built footprint and the city, offering a model for designing new corporate workspaces. The structure features terraces of varying dimensions, enhancing the spatial quality of renewing the stylistic expression of the corporate workspace (Fig. 2.32).

Fig. 2.32 João Moura, São Paulo, Brazil (2009), and sustainable façade applications

Location	São Paulo, Brazil
Köppen Climate Classification	Humid Subtropical Climate (Köppen: Cfa)
Completion Date	2009
Building Owner/Primary Tenant	Idea Zarvos
Architects & Design Teams	Nitsche Arquitetos
Number of Floors	10
Gross Floor Areas	6,300 m^2 including parking

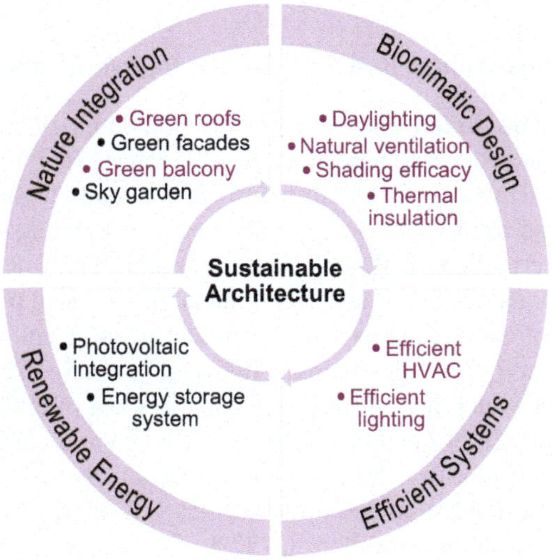

(4) **UTEC, Barranco, Peru, 2015**

Barranco, with a population of 48,000 inhabitants, is an upper-class coastal district of Lima, Peru, founded in 1872 [169]. The area features a rich array of architecture, ranging from colonial houses to Art Deco style; it was originally developed for wealthy families as a holiday destination and is now well known for its fashionable atmosphere, cultural life, beaches, museums, and churches [170]. Sustainable urban agriculture projects have reclaimed abandoned areas and transformed them into green spaces that contribute to improving food security and long-term food sustainability [171]. Local food programs promote food security within local communities, especially empowering women, by enabling agricultural practices within their neighborhoods. This helps to plant beneficial crops that enhance the urban landscape (Figs. 2.33 and 2.34).

As a district in Lima with a hot desert climate (Köppen: BWh), Barranco experiences warm weather throughout the year [172]. February is the hottest month of the year at approximately 23.2 °C (73.8 °F), whereas August is the coldest month of the year at approximately 16.5 °C (61.7 °F). Rainfall is relatively low, with the annual average of about 230 mm (9.1 in), which contributed to Barranco's dry climate. Winter months experience a fair amount of cloud cover and humid conditions.

Fig. 2.33 Exterior view of UTEC in Barranco, Peru (drawings reproduced courtesy of Grafton Architects and Shell Arquitectos)

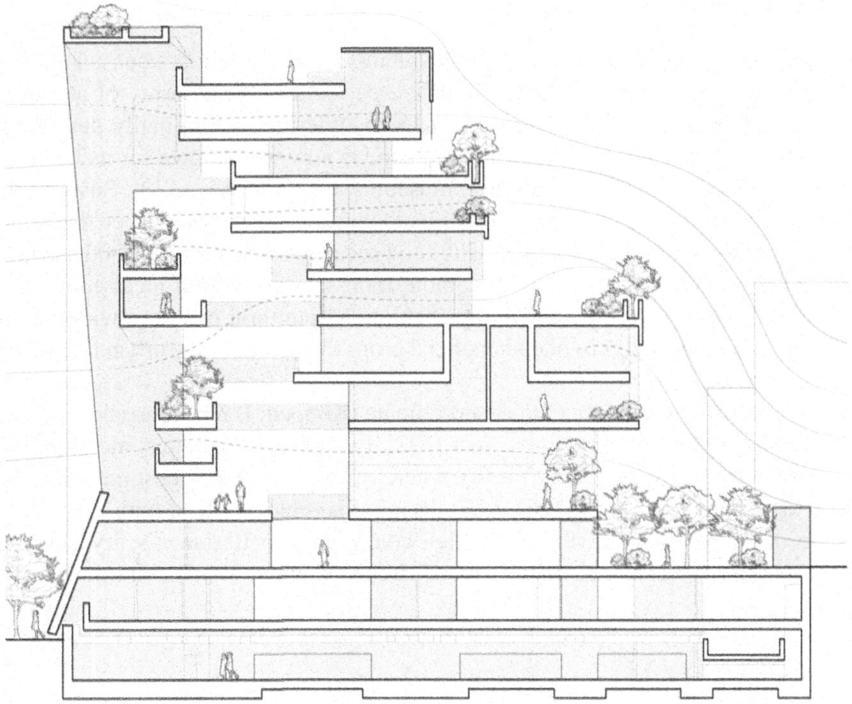

Fig. 2.34 Building section view of UTEC in Barranco, Peru (drawings reproduced courtesy of Grafton Architects and Shell Arquitectos)

The University of Engineering and Technology (UTEC) in Lima, Peru, is a notable example of modern sustainable architecture [173]. Designed by Shell Arquitectos and Grafton Architects, it combines new technologies with climate-responsive design strategies. According to the architects, the terraced landscaping of the building creates an amphitheater arrangement of the outdoor spaces suitable for gatherings while providing shade from the direct sunlight, particularly during hot periods. In addition, the design strategy helps minimize runoff caused by heavy rainfall, while simultaneously creating green areas suitable for relaxation or study within the university. Moreover, rainwater collection systems installed in various parts of the campus recycle graywater into the irrigation systems, efficiently watering landscapes and plants without wasting portable water. The building maximizes available daylight during normal working hours and reduces unnecessary electrical lighting consumption when classes are not in session. This sustainable design enables students to receive a high-quality education without compromising the environmental and aesthetic qualities of a traditional university campus (Fig. 2.35).

Fig. 2.35 UTEC, Barranco, Peru (2015), and sustainable façade applications

Location	Barranco, Peru
Köppen Climate Classification	Hot Desert Climate (Köppen: BWh)
Completion Date	2015
Building Owner/Primary Tenant	Universidad de Ingeniería y Tecnología (UTEC)
Architects & Design Teams	Grafton Architects, Shell Arquitectos
Number of Floors	10
Gross Floor Areas	68,624 m^2

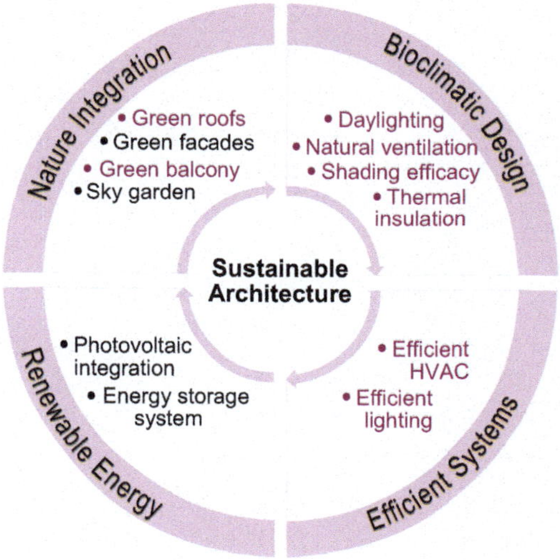

(5) **RB12, Rio de Janeiro, Brazil, 2016**

Rio de Janeiro is one of the largest South American cities with over 13.8 million people as of 2024, occupying a total area of 1221 km^2 (471.4 mi^2) [174]. Despite its beauty and iconic landmarks, Rio encounters issues such as poverty, crime, and environmental concerns [175]. These problems have recently been addressed through investment in public transport and enhanced law enforcement, among other initiatives, but significant challenges remain if the city aims to establish itself as a sustainable city. Climatic patterns in Rio de Janeiro are characterized by a tropical savanna climate (Köppen: Aw), which exhibits distinctive wet and dry seasons [176]. The climate is consistently warm, averaging about 23.6 °C (74.5 °F). Annual rainfall is about 1252 mm (49.3 in), with the most precipitation occurring during the wet season, which lasts from December to March, while very little rainfall occurs during the dry season from June to August. The hottest month of the year is February, with an average temperature of approximately 27 °C (80.6 °F). The coolest month is July, with mean temperatures of approximately 20.1 °C (68.3 °F). These climatic conditions help sustain a strong outdoor culture, especially during summer (Fig. 2.36).

Fig. 2.36 Exterior view of RB12 in Rio de Janeiro, Brazil (drawings reproduced courtesy of Victor Delaqua)

RB12 is a sustainable development project centered around renewable energy production, situated on Avenida Rio Branco within Rio de Janeiro's central business district [177]. The project involved the green refurbishment project of an existing building to meet sustainability requirements. Building data indicate that the innovative system improves water consumption, natural lighting, and natural ventilation, thereby increasing ecological sustainability and human well-being. RB12 contains diamond-shaped windows along its bioclimatic façades. It is also Brazil's first commercial building using photovoltaic panels for generating electricity that can be sold back into the grid and partially offset its energy demand. RB12 includes space for future fuel cell installation, allowing the building to convert street gas into electricity in future, although on-site electricity generation from gas is currently illegal under Brazilian regulations. The roof garden, green terrace and vegetated interiors help cool and moderate thermal conditions within the structure (Fig. 2.37).

Fig. 2.37 RB12, Rio De Janeiro, Brazil (2016), and sustainable façade features

Location	Rio De Janeiro, Brazil
Köppen Climate Classification	Tropical Savanna Climate (Köppen: Aw)
Completion Date	2016
Building Owner/Primary Tenant	Natekko
Architects & Design Teams	Triptyque Architecture
Number of Floors	21
Gross Floor Areas	4,728 m^2

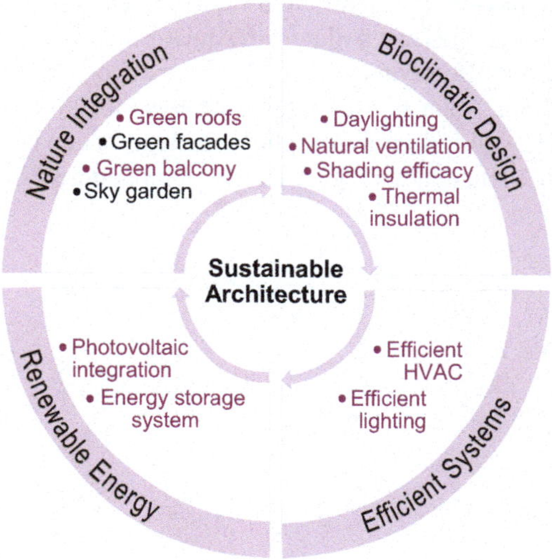

(6) Empresa de Desarrollo Urbano HQ, Medellin, Colombia, 2016

Medellin, Colombia's second-largest city, has approximately 4.1 million residents as of 2024, within an area of 380 km^2 (146.7 mi^2) [178]. Medellin's proximity to Caribbean ports and gold mines historically established its importance in South America, and today the city balances manufacturing with cultural attractions and urban development, including the expansion of its metro system [179]. Social urbanism is a strategy implemented by the municipality to address social inequality through urban planning, which includes public libraries, parks, schools, and public spaces that promote community interaction and social coherence [180]. Medellín, Colombia, has a temperate oceanic climate (Köppen: Cfb) characterized by mild temperatures year-round and evenly spread rainfall [181]. Medellín has a climate with heavy rainfall year-round at approximately 15.8 °C (60.4 °F) and about 2958 mm (116.5 in) of annual precipitation. Rain often comes in short, intense bursts, typically in the afternoon or night. Despite frequent rainfall, the city receives about 2555 h of sunshine annually.

EDU's new headquarters, a 37 m (121.4 ft) tall building covering 2983 m^2 (32,100 ft^2), is being developed through public-private collaboration [182]. According to the architect, the project aims to revitalize the city center by creating a safe, mixed-use area for citizens and transforming the center into a sustainable habitat, promoting social urbanism and sustainability. The design concept uses simple materials and intelligent geometries to create a "breathing building." An external skin directs cooler air to an internal solar chimney, enhancing thermal control and creating constant airflow in workspaces. The building incorporates an environmental chimney connecting office floors, drawing fresh air through windows and across floor plates. According to ventilation studies, each floor has three evenly-spaced openable windows on the least polluted and quietest facades [183]. The research suggests that during hot afternoons, the building uses two strategies: (1) a west-facing chimney provides a "solar boost", increasing fresh air flow by up to 33%, and (2) exposed concrete ceilings cool at night, absorbing occupants' radiant heat during the day, helping to maintain a cooler interior temperature (Fig. 2.38).

Location	Medellín, Antioquia, Colombia
Köppen Climate Classification	Oceanic Climate (Köppen: Cfb)
Completion Date	2016
Building Owner/Primary Tenant	Empresa de Desarrollo Urbano (EDU)
Architects & Design Teams	EDU (project team), with collaborators including Salmaan Craig and technical consultant
Number of Floors	10
Gross Floor Areas	3,660 m2

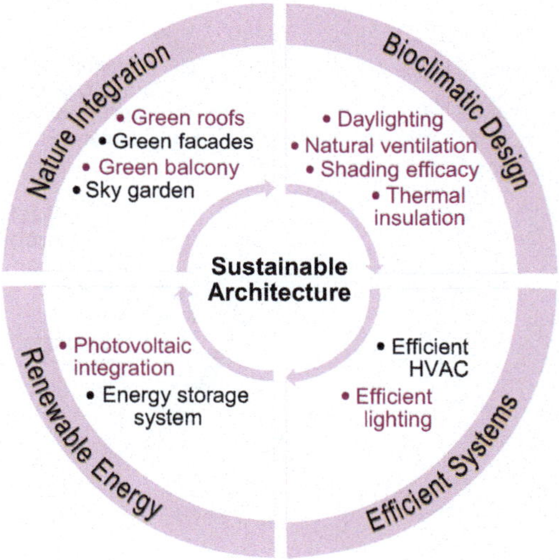

Fig. 2.38 Empresa de Desarrollo Urbano, Medellin, Colombia (2016), and sustainable façade applications

(7) Hippodrome Tower, Zapopan, Mexico, 2018

The Hippodrome Tower in Guadalajara, Mexico (2012–18), designed by OAB with Carlos Ferrater and Xavier Martí, is a landmark achievement, that rises amidst surviving pre-urban streets, addressing the area's lack of planned infrastructure [184]. The design intent was to create a 13-story irregular prism that brings order to a chaotic growth area of Jalisco's capital, where diverse building types and inadequate road systems coexist. According to the architect, the new building features a complex geometric grid on an irregular plot, fan-shaped floor slabs, a stepped entrance with a double-height ground floor, a central space inspired by Mies van der Rohe's Seagram Building, and a patio with space, light, and vegetation. Building data indicate that the facade openings incorporate six modular perforated panels with a central opening, filtering light, solar radiation, and airflow. In addition, the lightweight, prefabricated

Location	Zapopan, Mexico
Köppen Climate Classification	Hot-Summer Mediterranean Climate (Köppen: Csa)
Completion Date	2018
Building Owner/Primary Tenant	Grupo VEQ
Architects & Design Teams	OAB (Office of Architecture in Barcelona)
Number of Floors	13
Gross Floor Areas	5,267 m^2

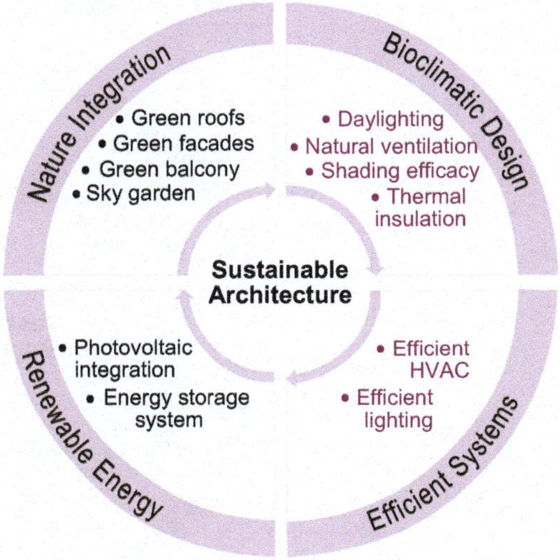

Fig. 2.39 Hippodrome Tower, Zapopan, Mexico (2018), and sustainable façade applications

glass fiber-reinforced concrete (GRC) screen serves as an anti-seismic elements and metaphorical keystones, enabling assembly from modulated openings to the tower's stepped, pivoted form through easy insertion of the panels. The view from inside the building is framed by modulated openings, while, at night, the tower becomes a lantern, its geometrically perforated skin emitting light and creating a pixelated image (Fig. 2.39).

(8) **Universidad EAN, Bogotá, Columbia, 2019**

The capital city of Colombia, Bogotá, one of the highest capital cities in South America, with its total population of approximately 11.6 million as of 2024, is situated on the Andean Cordillera, in the Bogotá savanna [186]. The city has an area of 1587 km^2 (226.6 mi^2), is located at approximately 2640 m (8661 ft) above sea level, and is organized in a grid plan that features notable plazas such as Plaza Bolívar. Urban planning strategies in Bogotá include an integrated approach combining the improvement of public transportation, the enhancement of environmental sustainability in construction, and the provision of public safety to raise urban living standards [187]. The city is at a vegetation zone that suggests it falls under a warm-summer Mediterranean climate (Köppen: Csb), characterized by warm dry summers and mild wet winters [188]. Potential climate extremes are moderated by the city's high altitude. The average temperature is approximately 12.9 °C (55.2 °F). The average annual rainfall is around 1090 mm (42.9 in), distributed evenly throughout the year, with the rainiest months being April–May and October–November. The city maintains a temperate, cool, and humid climate through the seasons (Fig. 2.40).

Universidad EAN has built its first campus building designed with Cradle-to-Cradle principles, focusing on environmental sustainability and energy efficiency through circular approaches [189]. According to the architect, the design ensures that all architectural systems and materials in the building are intended for reuse after their life cycle ends. Solar heat gain or loss through windows is managed through strategic window placement coupled with shading devices. Insulation foam filled within concrete blocks reduces energy flow and maximizes energy savings in opaque walls. A vegetative roof effectively moderates roof temperatures during the summer, while also providing additional protection from external noise. Together, these features contribute to the university's sustainability goals, compared to other conventional design approaches in which environmental and social benefits are often not considered at the planning stage (Fig. 2.41).

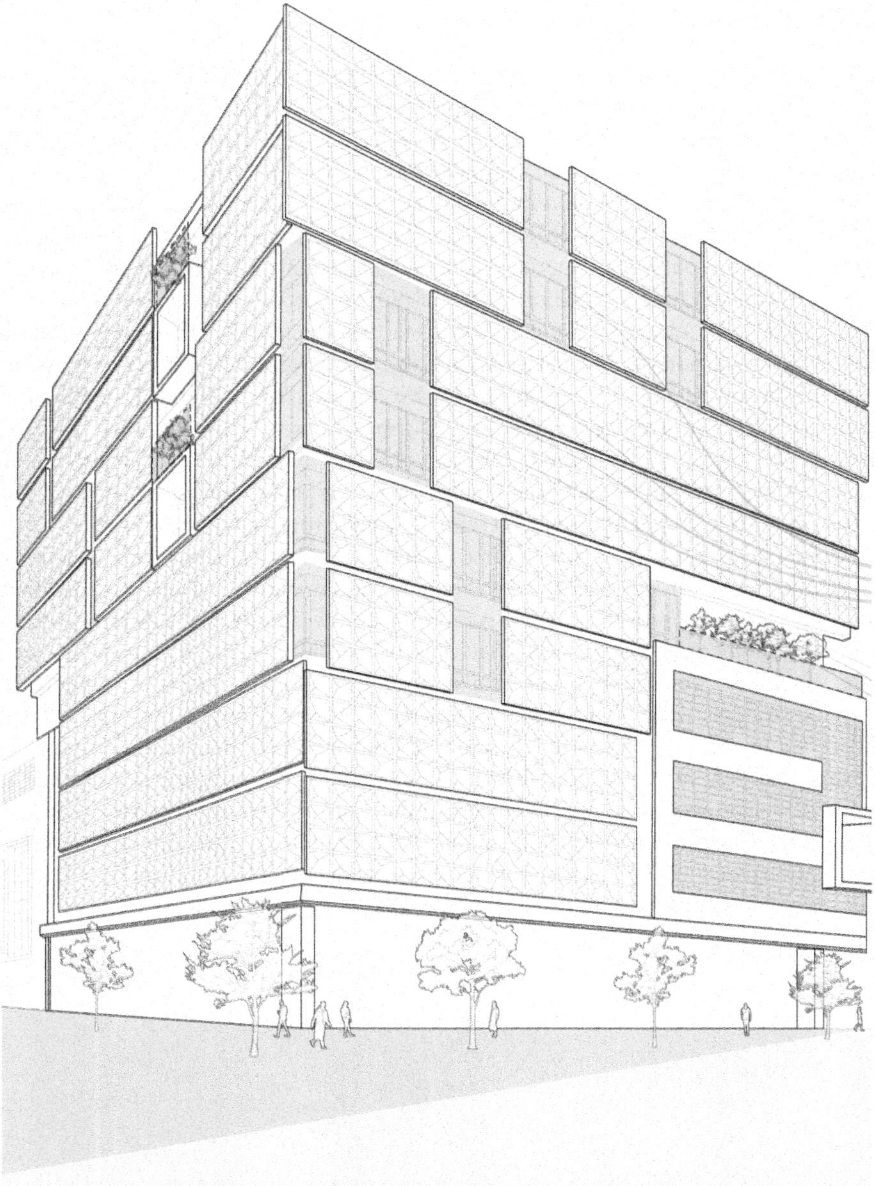

Fig. 2.40 Exterior view of Universidad EAN in Bogotá, Colombia (drawings reproduced courtesy of William McDonough + partners)

Fig. 2.41 Universidad EAN, Bogotá, Colombia (2019), and sustainable façade applications

Location	Bogotá, Colombia
Köppen Climate Classification	Warm-summer Mediterranean Climate (Köppen: Csb)
Completion Date	2019
Building Owner/Primary Tenant	Educational University (Universidad EAN)
Architects & Design Teams	William McDonough + Partners
Number of Floors	6
Gross Floor Areas	20,000 m^2

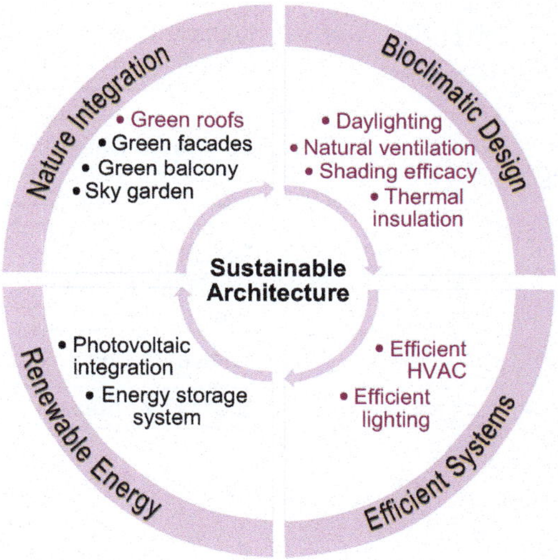

(9) Complejo Atrio, Bogotá, Colombia, 2024

Atrio WSP Colombia is a leading water and sanitation service provider in Colombia, and handles urban wastewater control, solid waste collection, and industrial effluent treatment for government and private clients [190]. The Atrio project is a major mixed-use development in central Bogotá featuring two towers and a large public space [191]. The North (201 m; 659.4 ft) and South (268 m; 879.3 ft) towers provide over 250,000 m^2 (2,691,000 ft^2) of office space, public services, and retail, accommodating up to 72,000 daily visitors. With an additional 10,000 m^2 (107,639 ft^2) of public space, the project aims to revitalize the Centro Internacional area.

The building has taken steps to enhance its environmental sustainability as well as reduce operational costs [192]. The curtainwall façade system efficiently controls solar heat gains whilst ensuring optimal daylighting requirements are met. The building integrates energy-efficient environmental control systems and uses recyclable construction materials. The project team participates in reforestation programs using native species near Bogotá and events such as World Water Day to advocate for access to clean water. These activities underline the company's determination to engage in sustainable development and build strong relationships with customers and partners in Colombia (Fig. 2.42).

Fig. 2.42 Complejo Atrio, Bogotá, Colombia (2024), and sustainable façade applications

Location	Bogotá , Colombia
Köppen Climate Classification	Warm-Summer Mediterranean Climate (Köppen: Csb)
Completion Date	North Tower Completed in 2019, South Tower Under Construction (2024)
Building Owner/Primary Tenant	Office
Architects & Design Teams	Rogers Stirk Harbour + Partners
Number of Floors	44 (North Tower), 67 (South Tower)
Gross Floor Areas	239,000 m^2

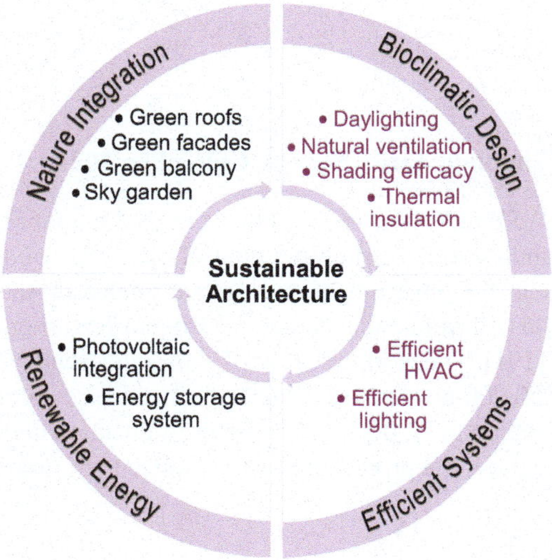

2.2.3 *Tall Building Enclosures in Hot Climates: African*

(1) Eastgate Center, Harare, Zimbabwe, 1995

Harare is the capital of Zimbabwe with a population of approximately 1.6 million as of 2024, and covers a land area of 960 km^2 (370.7 mi^2) [193]. A significant portion of the population consists of migrant workers from neighboring countries, who come for employment and business purposes [194]. The city is a hub of manufacturing, agriculture, tourism, and finance, and acts as an international trade point as it is situated between Mozambique and South Africa [195]. Harare is classified under subtropical highland climate (Köppen: Cwb), with mild dry, winters and warm, wet summers throughout the year [196]. As is common in many tropical highland regions, warm rainy seasons occur during November to March while cool, dry seasons occur from May to August, during which nights are cooler than at other times of the year (Figs. 2.43, 2.44).

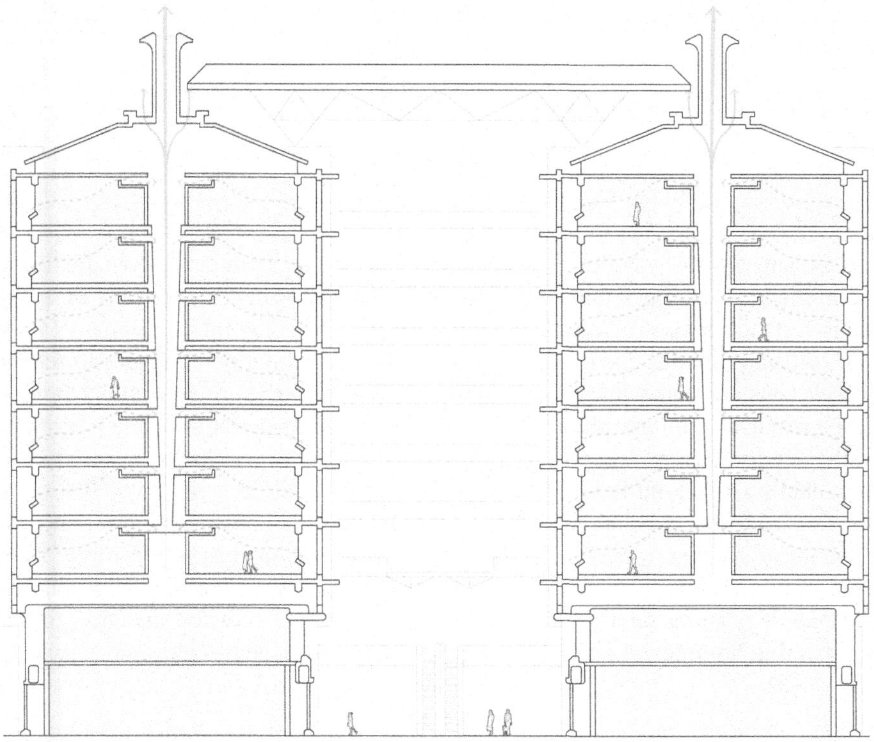

Fig. 2.43 Exterior view of Eastgate Centre in Zimbabwe (drawings reproduced courtesy of Mick Pearce Architect)

Fig. 2.44 Building Section View of Eastgate Centre in Zimbabwe (drawings reproduced from Mick Pearce Architect)

Eastgate Centre was designed by Mick Pearce, an architect known for using biomimicry as an inspiration for design in Zimbabwe and completed in 1996; today it stands out one of among the most energy-efficient buildings globally [197]. The building was designed primarily using natural ventilation strategies to reduce energy consumption while maintaining indoor comfort levels at optimal ranges. Large windows are tactically placed to allow sunlight penetration while perforated walls increase airflow around the complex, and each floor includes louvered apertures that provide ventilation control. This unique façade design and spatial organization allow air circulation to occur naturally indoors, thus reducing reliance on mechanical cooling systems and leading to significant savings in electricity bills. Many sustainability-leading architects increasingly adopt biomimicry techniques because of benefits such as increased thermal comfort, reduced operating costs, improved environmental sustainability, and enhanced occupant health and well-being (Fig. 2.45).

Fig. 2.45 Eastgate Centre, Harare, Zimbabwe (1996), and sustainable façade applications

Location	Harare, Zimbabwe
Köppen Climate Classification	Subtropical Highland Climate (Köppen: Cwb)
Completion Date	1996
Building Owner/Primary Tenant	Old Mutual Zimbabwe
Architects & Design Teams	Mick Pearce
Number of Floors	7
Gross Floor Areas	32,000 m^2

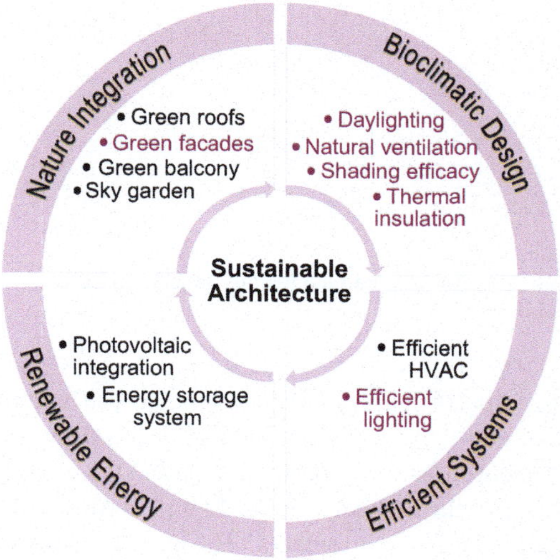

(2) One Airport Square, Accra, Ghana, 2015

Accra, Ghana's capital, has a population of approximately 2.7 million as of 2024 and covers a land area of about 225 km^2 (86.9 mi^2) [198]. Its strategic location between Europe and Africa fostered early growth, initially through fishing industries, and Accra is continuing to advance urban development through sustainability initiatives [199]. The 2018 Greening and Beautification Project engages businesses in improving public green spaces, air quality, flood prevention, and biodiversity [200]. The initiative focuses on urban agriculture and sustainable building to enhance residents' quality of life. Accra has a hot semi-arid climate (Köppen: BSh) [201]. The wet season (April–October) brings heavy and frequent rainfall, with peaks in May–June. The dry season (November–March) is characterized by lower humidity, little rainfall, and warm temperatures year-round (Fig. 2.46).

Fig. 2.46 Exterior view of One Airport Square in Ghana (drawings reproduced courtesy of Mario Cucinella)

One Airport Square is a ten-story mixed-use building with 17,000 m^2 (182,986 ft^2) of office and commercial space and inspired by local art and palm tree bark; its design incorporates environmental strategies to address climate challenges [202]. According to the architect, the building's form balances functional needs with structural and energy requirements. The project earned 4-Star rating from the Green Building Council of South Africa, and its design, based on site analysis and optimal orientation, sets a standard for West African office buildings. The project addresses high solar exposure and year-round warmth by implementing strategies to reduce cooling needs. The building optimizes the use of natural light and fresh air, which are uniformly distributed throughout the interior spaces via a raised floor system. To reduce portable water consumption, technologies are employed to limit flow rates, the usage duration, and delivery timing, while graywater and rainwater are reused for toilet flushing. Varying shaft depths create three office types: maxi, midi, and mini. Projecting terraces, which recede as they rise, define the building's dynamic form. These variations allow flexible spatial configurations to accommodate diverse tenant needs (Fig. 2.47).

Fig. 2.47 One Airport Square, Accra, Ghana (2015), and sustainable façade applications

Location	Accra, Ghana
Köppen Climate Classification	Hot Semi-Arid Climate (Köppen: BSh)
Completion Date	2015
Building Owner/Primary Tenant	Eni Ghana Exploration & Production
Architects & Design Teams	Mario Cucinella Architects
Number of Floors	9
Gross Floor Areas	17,000 m^2

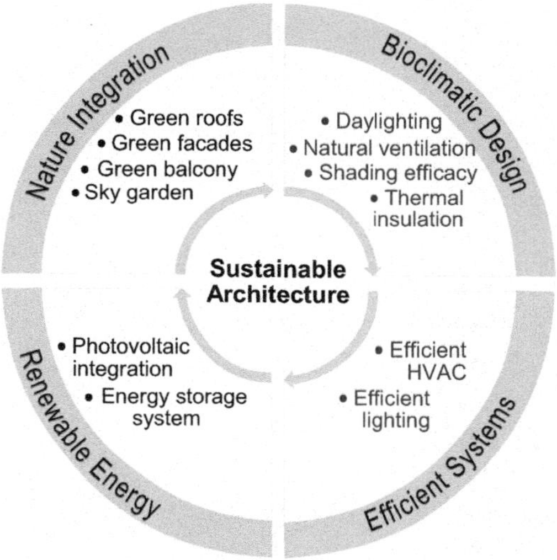

(3) **Lideta Mercato, Addis Ababa, Ethiopia, 2016**

Addis Ababa, Ethiopia's capital city, has a population of approximately 5.7 million people as of 2024, covering an area of 527 km^2. The city is located in the Central region of the country, surrounded by hills and mountains on a well-watered plateau with an elevation of approximately 2355 m (7726 ft) [203]. Addis Ababa is becoming a modern resilient city. For example, the City Corridor Development Project restores abandoned areas with bike lanes, footpaths, gardens, libraries, and houses [204]. Likewise, the city's Integrated Housing and Development Program offers affordable housing and increases homeownership rates by investing heavily in the construction of condominium apartments. These urban development projects collectively aim to improve infrastructure, enhance public services, and foster economic growth, ultimately striving to make Addis Ababa a more livable and efficient metropolis. Addis Ababa, Ethiopia, has an oceanic climate (Köppen: Cfb) characterized by mild temperatures year-round and evenly spread rainfall [205]. Mild, dry winters typically last from October to February, with cooler temperatures, especially at night. Moderate, wet summers occur from June to September, with the majority of the annual rainfall of approximately 1874 mm (73.8 in) falling during these months (Figs. 2.48 and 2.49).

Fig. 2.48 Exterior view of Lideta Mercato in Addis Ababa, Ethiopia (drawings reproduced courtesy of Vilalta Studio)

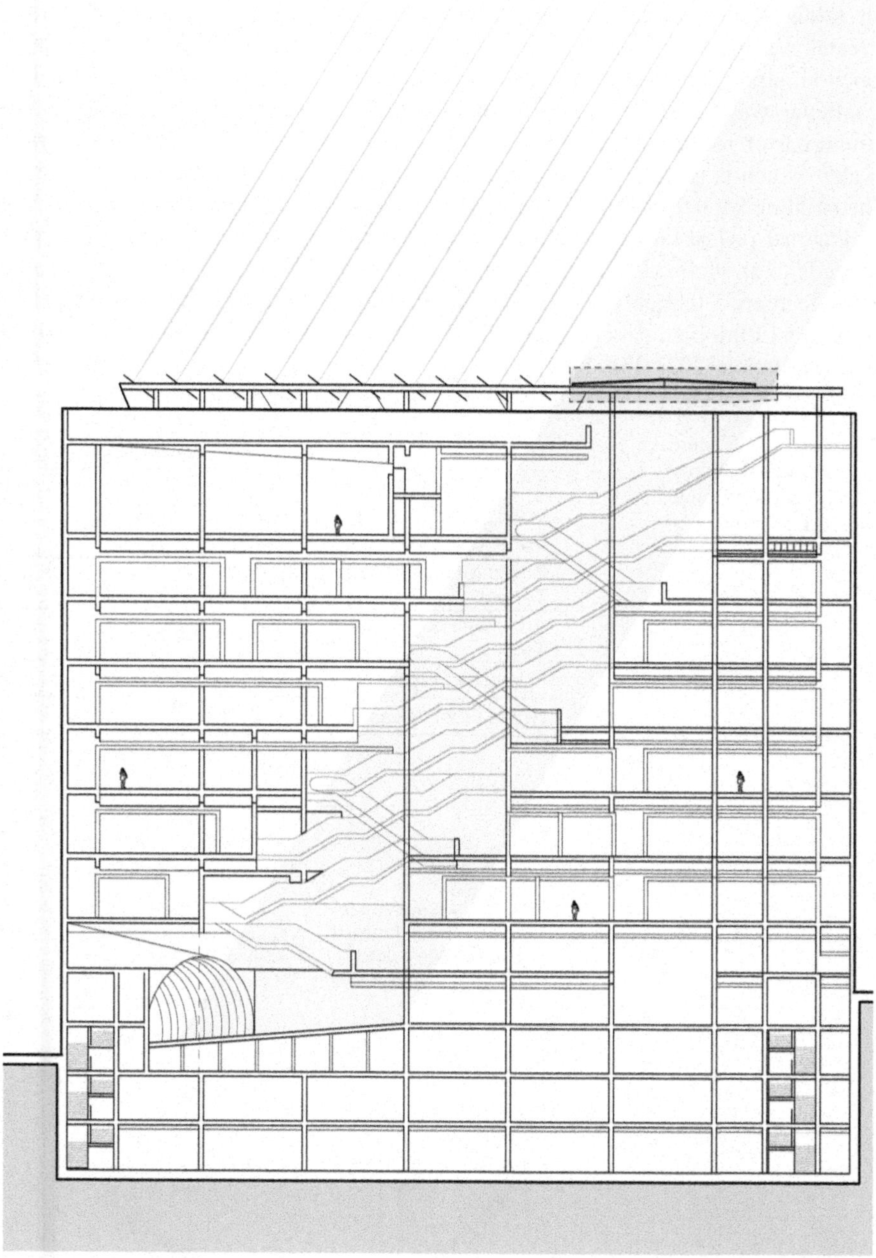

Fig. 2.49 Building section view of Lideta Mercato in Addis Ababa, Ethiopia (drawings reproduced courtesy of Vilalta Studio)

The Lideta Mercato, inspired by the Old Mercato, was designed as a multi-story market to address thermal and lighting issues commonly found in conventional malls in Addis Ababa [206]. This design, which considers local climate and traditions, creates a natural ventilation system and balanced interior lighting, resulting in an open-air atmosphere within the atrium. Ethiopia faces frequent power outages. This challenge was transformed into an opportunity by installing circular photovoltaic umbrellas on rooftops, creating shaded entertainment areas. The roof collects rainwater, which is stored in basement tanks. After filtration, this water is reused for the building's toilets and taps. The carved volume links two parallel streets, creating a diagonal pedestrian shortcut and concentrates all building entrances. This path extends to an inclined central atrium connecting all floors. Around this void, each floor features a unique layout of small shops. The building's facade, inspired by traditional Ethiopian patterns, uses lightweight concrete prefabs to control sunlight and ventilation [207] (Fig. 2.50).

Fig. 2.50 Lideta Mercato, Addis Ababa, Ethiopia (2016), and sustainable façade applications

Location	Addis Ababa, Ethiopia
Köppen Climate Classification	Oceanic Climate (Köppen: Cfb)
Completion Date	2016
Building Owner/Primary Tenant	Addis Ababa City Government
Architects & Design Teams	Vilalta Studio
Number of Floors	8
Gross Floor Areas	14,200 m^2

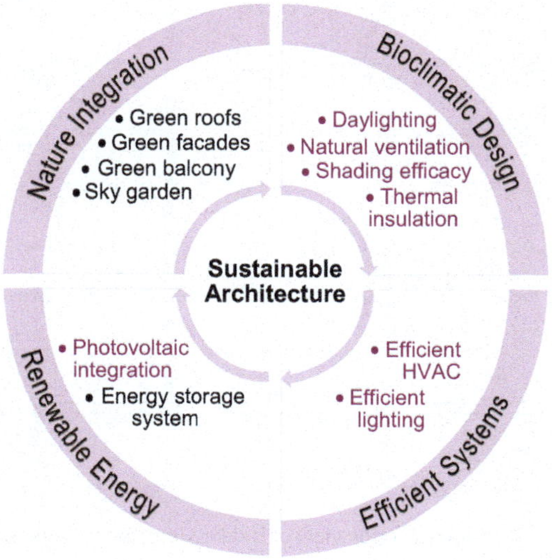

(4) **Kingsway Tower, Lagos, Nigeria, 2019**

Lagos, the largest city and economic heart of Nigeria, is among the most highly populated cities in Africa and one of its fastest-growing ones with more than 16 million inhabitants as of 2024, within a city area of about 1171 km^2 (452.1 mi^2) [208]. Research indicates that approximately two-thirds of the city's population live in slums, while most developable land is urbanized [209]. Land registration status affects poverty and the environment, and insecure land acquisition deters investment. These factors contribute to the persistence of poverty in informal settlements. Lagos, Nigeria has a tropical monsoon climate (Köppen: Am) characterized by short dry season and intense monsoon rains [210]. Lagos has a warm climate year-round, with temperatures averaging approximately 26.7 °C (80 °F). The city experiences a wet season from April to October, peaking in July and August, and a dry season from November to March. The annual rainfall is about 1783 mm (70.2 in) (Figs. 2.51 and 2.52).

The Kingsway Tower located in Lagos is a 15-story mixed-use tower containing retail spaces, offices, and underground parking [211]. Its sail-inspired form reflects the city's maritime heritage. The long facades of the building are oriented along the north-south axis to reduce solar stress caused by the high sun angle. This building orientation, coupled with shading devices, effectively manages climate impact. The retail podium is accentuated by a wavy canopy that shields pedestrians from harsh outdoor environments while articulating the building's primary entry zone. The building façade resembles a storefront rather than a curtainwall, with the slab extending beyond the glass facades. Folded and perforated aluminum screens give the building's identity, drawing strong inspiration from tropical vegetation patterns. These screens provide a porous skin that creates internal shadows within the building. The aluminum screen is located at the slab's edge to further reduce the solar load and facilitate maintenance access. Lagos struggles with irregular power supply and poor municipal services. This landmark building serves as a Nigeria's economic icon. In response, the building integrates energy-efficient technologies, lowering air-conditioning needs and increasing daylight penetration. Additional building services, including water supply and treatment, emergency back-up power, and sewage services, are provided on the site (Fig. 2.53).

Fig. 2.51 Exterior view of Kingsway Tower in Lagos, Nigeria (drawings reproduced courtesy of SAOTA)

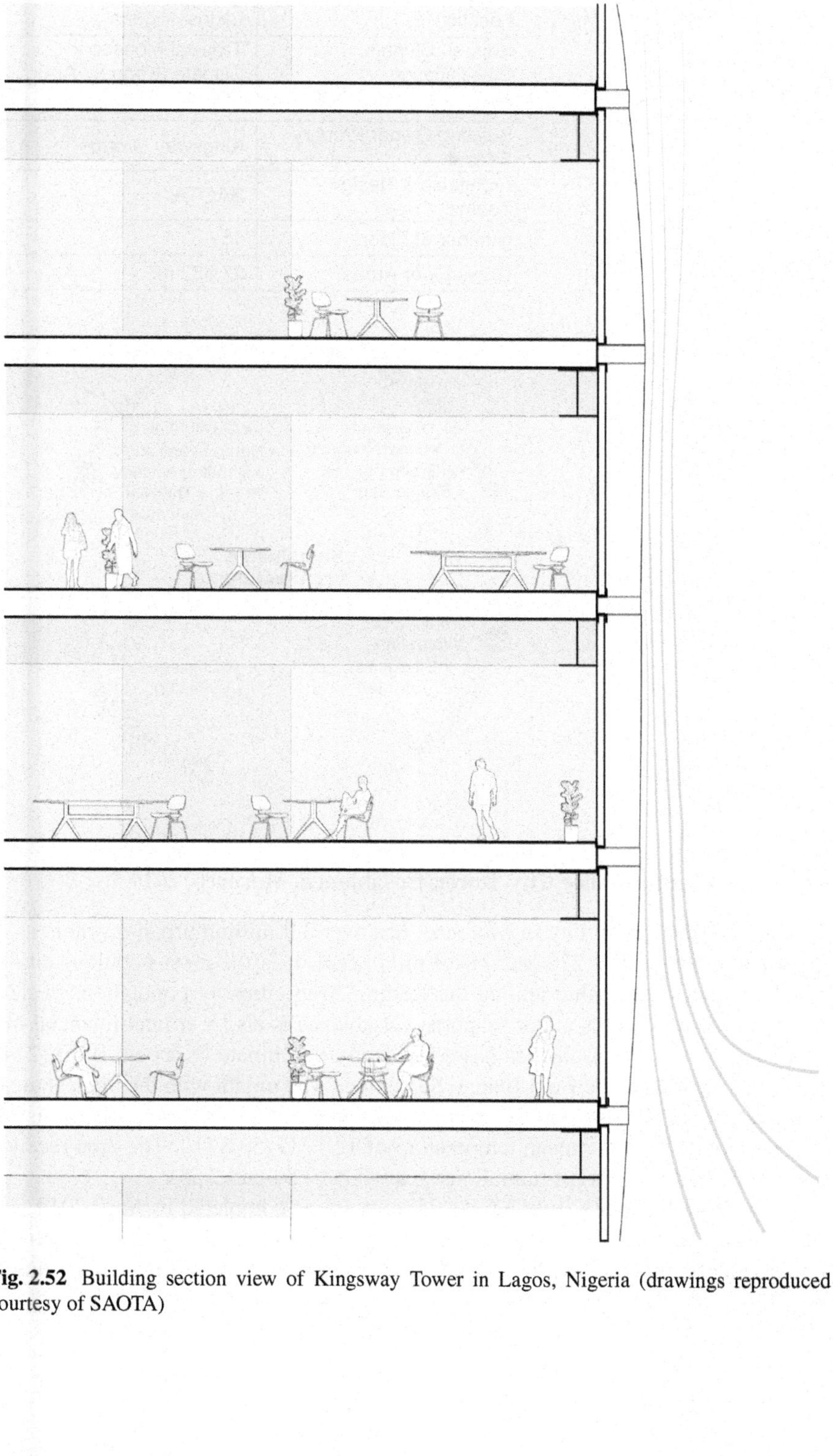

Fig. 2.52 Building section view of Kingsway Tower in Lagos, Nigeria (drawings reproduced courtesy of SAOTA)

Fig. 2.53 Kingsway Tower, Lagos, Nigeria (2019), and sustainable façade applications

Location	Lagos, Nigeria
Köppen Climate Classification	Tropical Monsoon Climate (Köppen: Am)
Completion Date	2019
Building Owner/Primary Tenant	Kingsway Group
Architects & Design Teams	SAOTA
Number of Floors	15
Gross Floor Areas	27,832 m^2

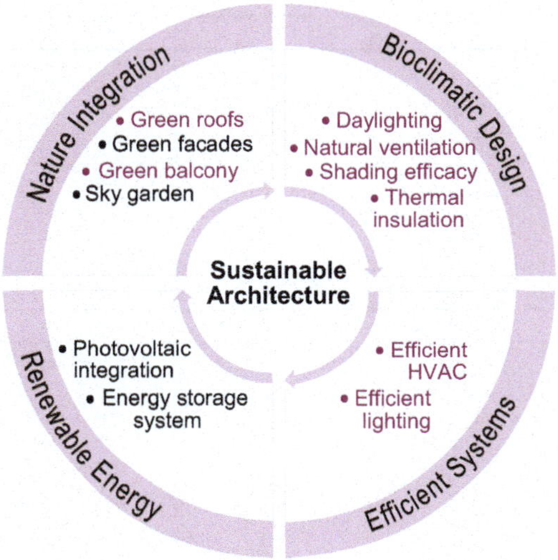

(5) Casablanca Finance City Tower, Casablanca, Morocco, 2019

Casablanca, the largest city in Morocco, has over 3.9 million urban dwellers as of 2024 and covers about 220 km^2 (84.9 mi^2) [212]. It is the most populous city in Maghreb and ranks eighth among the leading Arab cities by population. Besides being one of Morocco's major seaports, Casablanca is also a critical financial hub in the region [213]. Casablanca has a hot semi-arid climate (Köppen: BSh) [214]. Hot, dry summers range from June to September. The month with the highest mean temperatures is August, with an average of about 24 °C (75.2 °F). January is the coldest month, with the mean temperature of 12.1 °C (53.8 °F). The area receives nearly 324 mm (12.8 in) of rainfall per year (Figs. 2.54 and 2.55).

Casablanca Finance City (CFC), a 25-story tower completed in April 2019, is a new business district located southwest of downtown the Anfa District and attracts international companies establishing Moroccan regional offices within this economic

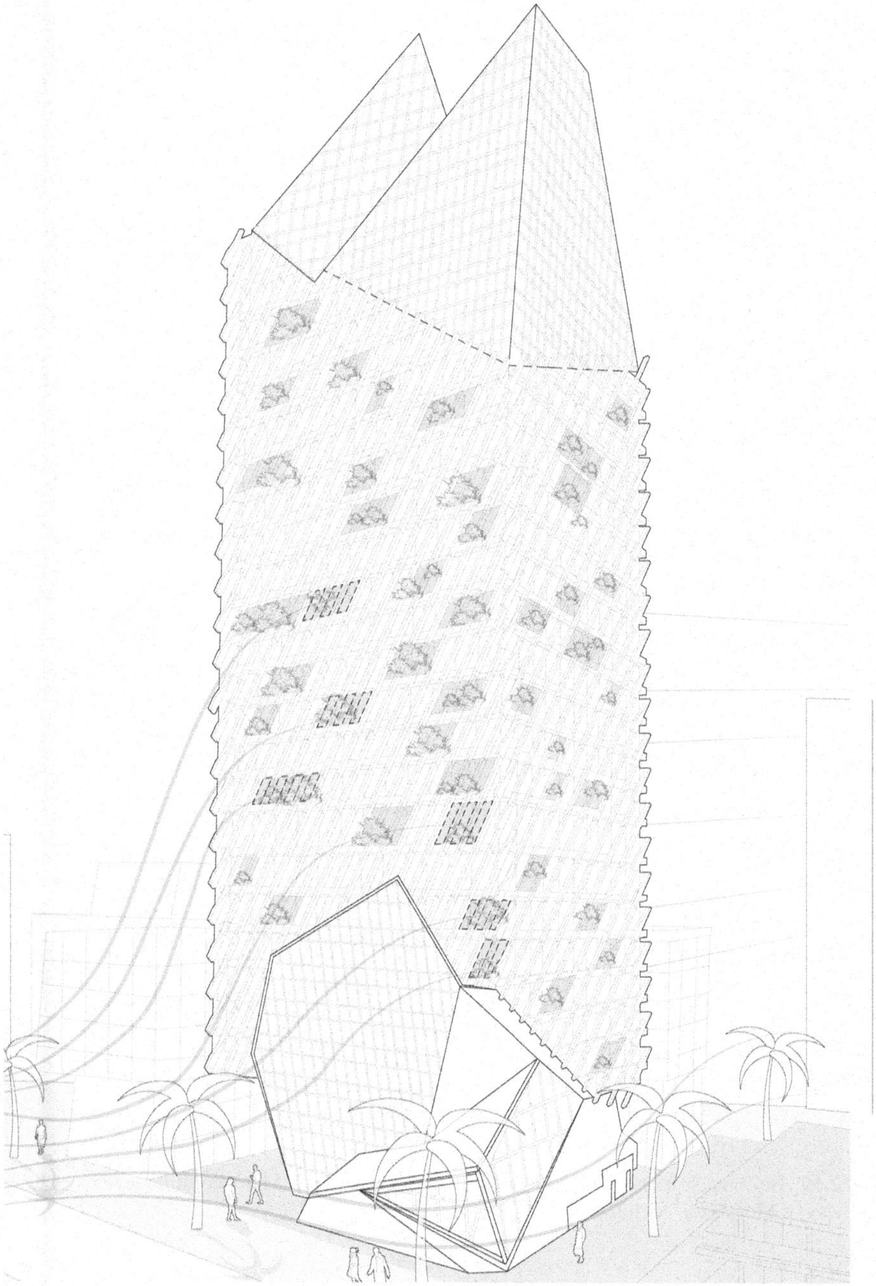

Fig. 2.54 Exterior view of Casablanca Finance Tower in Morocco (drawings reproduced courtesy of Morphosis Architects)

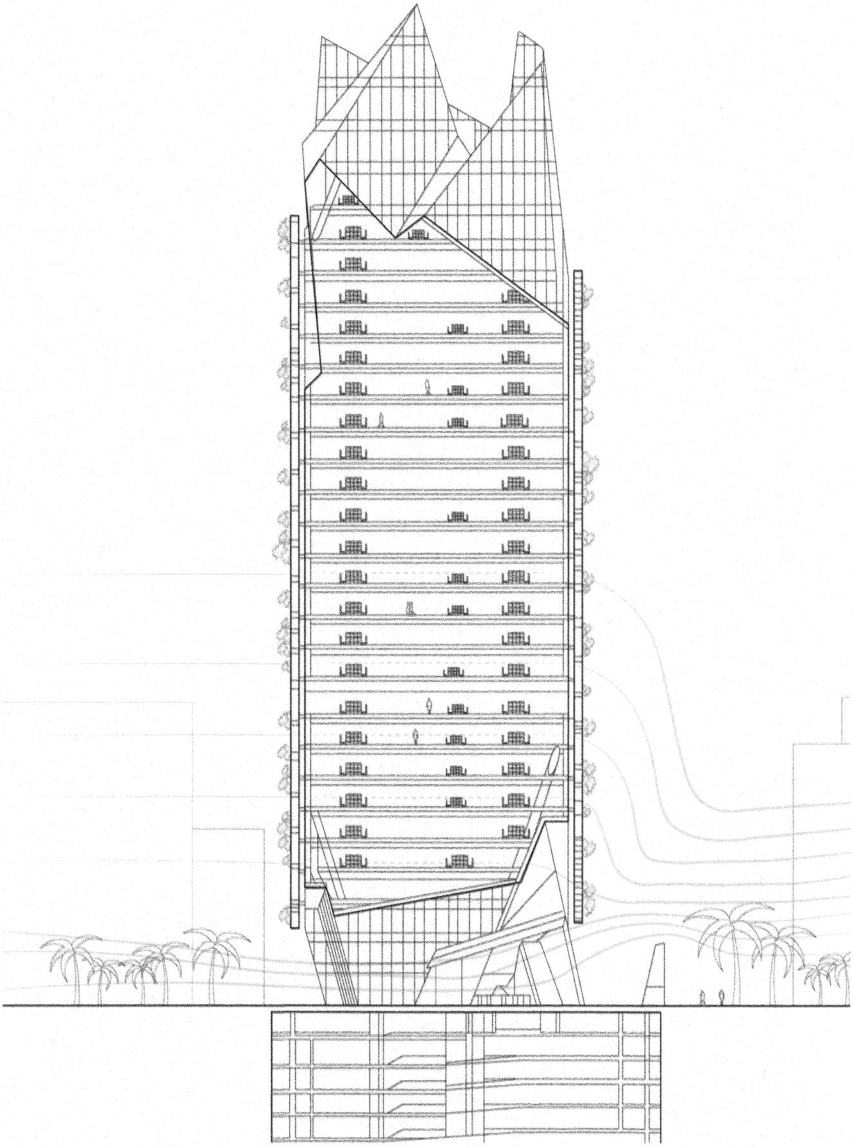

Fig. 2.55 Building section view of Casablanca Finance Tower in Morocco (drawings reproduced courtesy of Morphosis Architects)

zone [215]. Designed to adapt to local climatic conditions and minimize ecological impacts, the tower sets a new benchmark for regional and global projects. The exterior incorporates Moroccan-inspired geometric patterns alongside latticework. On the ground level, shaded outdoor spaces invite public participation. The design functions both as a metaphor for urban development and an active social hub, serving as a focal

Location	Casablanca, Morocco
Köppen Climate Classification	Hot Semi-Arid Climate (Köppen: BSh)
Completion Date	2019
Building Owner/Primary Tenant	Casablanca Finance City Authority
Architects & Design Teams	Morphosis
Number of Floors	25
Gross Floor Areas	21,000 m^2

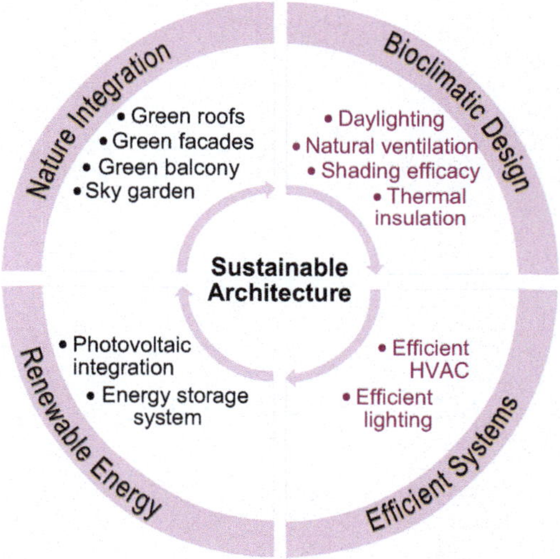

Fig. 2.56 Casablanca Finance Tower, Casablanca, Morocco (2019), and sustainable façade applications

point of street-level activity within the neighborhood. Building data indicate that the project achieved LEED Gold certification through energy efficient design. A brise-soleil system lets in sunlight and natural ventilation while also controlling solar exposure (Fig. 2.56).

(6) **The Bank, Johannesburg, South Africa, 2021**

Johannesburg is South Africa's largest city, with approximately 6.4 million residents as of 2024, and covering a land area of 334 km^2 (129 mi^2), located on the Highveld plateau at an elevation of 1753 m (5751 ft) above sea level [216]. The Making and Shaping of the City by Keith Beavon provides a detailed analysis of its transformation from a mining town to a modern metropolis [217]. This involves studying issues such as how gold mining created jobs and reinforced racial segregation, alongside factors such as transport infrastructure and immigration. Ongoing issues include urban renewal projects, smart city initiatives, and sustainability efforts. Johannesburg has a subtropical highland climate (Köppen: Cwb), characterized by dry winters, mild temperatures due to elevation [218]. January has an average temperature of approximately 19.7 °C (67.4 °F), making it the hottest month of the year. July experiences the lowest average temperature at approximately 9.5 °C (49.1 °F). The city receives about 784 mm (30.9 in) of annual rainfall, mostly during summer (October–April). Johannesburg experiences abundant sunshine, averaging 3334 h annually (Fig. 2.57).

Fig. 2.57 Exterior view of The Bank in Johannesburg, South Africa (drawings reproduced courtesy of DAFFONCHIO Architects + Imbewu Design)

The Bank at Rosebank is a 13-story 15,000 m^2 (161,459 ft^2) mixed-use development in Rosebank, Johannesburg, built on an existing four-story bank structure, blending modern design and historical preservation [219]. The building's black face-brick façade, softened by greenery, reflects its banking history. This material also contributes to reducing energy consumption by reflecting heat away from the interior spaces during hot summer months. Sustainable practices include LED lighting and energy efficient air conditioning systems, ensuring comfort in all weather conditions. The building features landscaped roof gardens overlooking Johannesburg. The Bank at Rosebank exemplifies adaptive reuse architecture, blending harmoniously with its urban surroundings while remaining aesthetically distinctive in the city (Fig. 2.58).

Fig. 2.58 The Bank, Johannesburg, South Africa (2021), and sustainable façade applications

Location	Johannesburg, South Africa
Köppen Climate Classification	Subtropical Highland Climate (Köppen: Cwb)
Completion Date	2021
Building Owner/Primary Tenant	Attacq Limited
Architects & Design Teams	Daffonchio Architects
Number of Floors	13
Gross Floor Areas	15,000 m^2

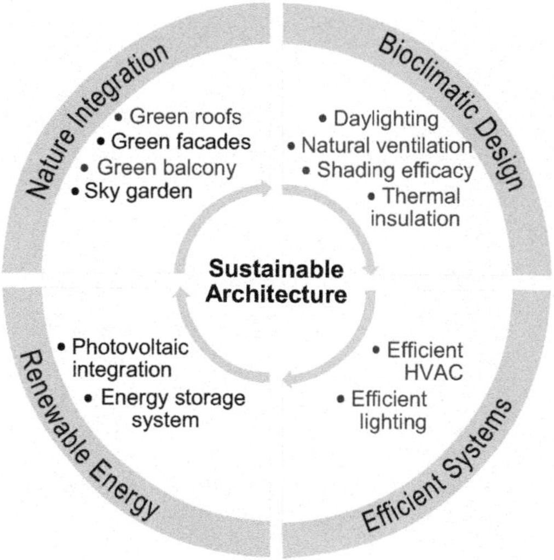

2.2.4 *Tall Building Enclosures in Hot Climates: Oceania*

CH2, Melbourne, Australia, 2006

Melbourne is the capital city of the state of Victoria and the second most populous city in Australia, with a population of approximately 5.3 million people as of 2024 and covering an extensive area of about 9993 km^2 (3858.3 mi^2) [220]. Melbourne's urban history is rich and complex, influenced by economic conditions, population changes, government policies, and cultural factors [221]. Davison notes that while some early planning attempts occurred, such as Light's Adelaide plan (1837–1838), most Australian cities grew organically due to rapid European immigration in the mid-nineteenth century, and Melbourne experienced significant growth from the 1850s to 1870s, becoming one of Australia's largest cities by the 1880s. Despite these changes, Australian cities have maintained much of their original character through preservation efforts. Melbourne has an oceanic climate (Köppen: Cfb), characterized by mild temperatures year-round and evenly spread rainfall [222]. The city experiences warm summers and cool winters. The peak of summer, during January and February, sees average temperatures around 20.2 °C (68.4 °F) and 20.1 °C (68.2 °F), respectively. In contrast, the winter chill is most pronounced in July, with an average temperature dropping to approximately 9.6 °C (49.3 °F). Rainfall averages about 645 mm (25.4 in) per year and is relatively evenly distributed throughout the year. The city is also famous for its changeable weather, often described as "four seasons in one day" (Figs. 2.59 and 2.60).

Fig. 2.59 Exterior view of CH2 building in Melbourne, Australia (drawings reproduced courtesy of Mick Pearce and DesignInc)

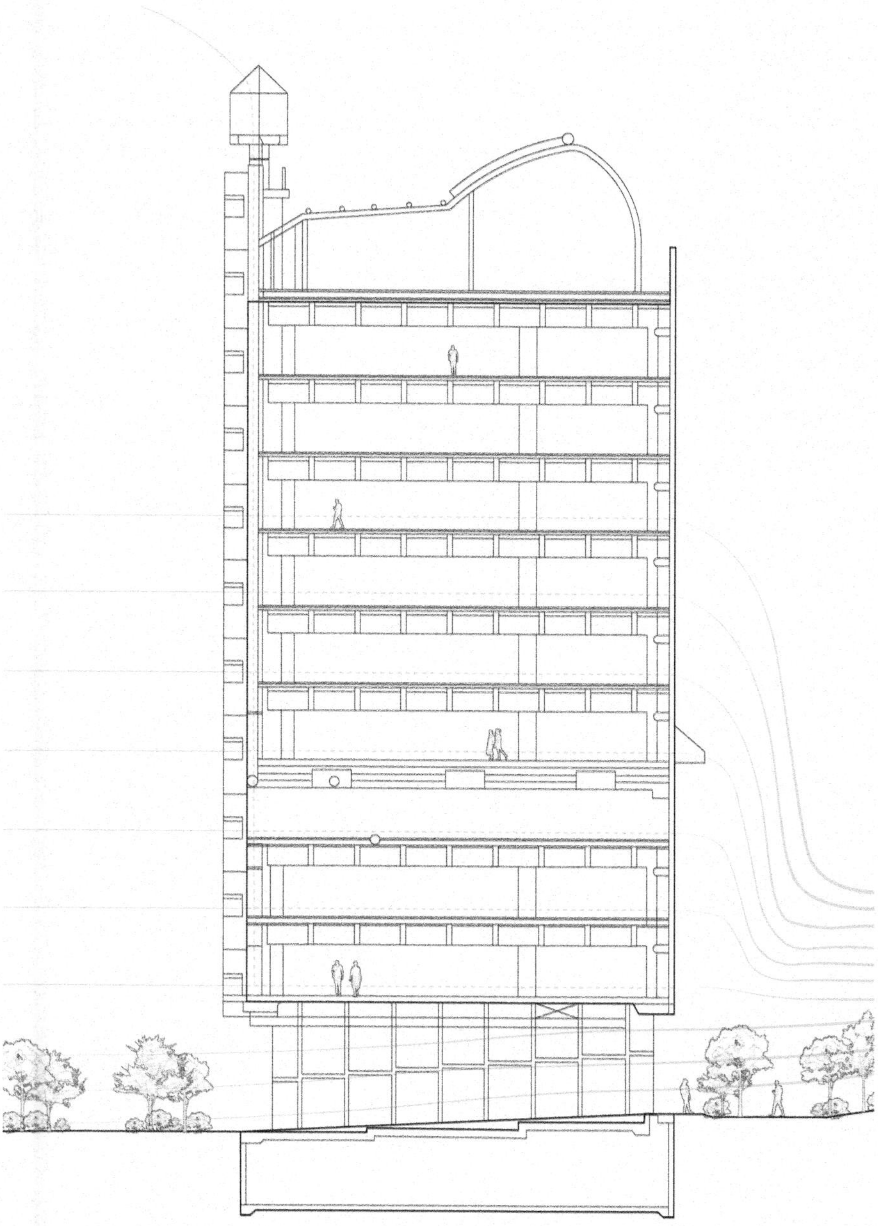

Fig. 2.60 Building section view of CH2 building in Melbourne, Australia (drawings reproduced courtesy of Mick Pearce and DesignInc)

The CH2 building in Melbourne, Australia, designed by Mick Pearce and DesignInc, is a landmark building mimicking natural processes [223]. This iconic office building incorporates biomimicry principles, with features such as a tree-branch-inspired roof designed for optimal light penetration. Energy-efficient building service systems and solar panels reduce emissions. The design prioritizes occupant comfort with operable windows for fresh air circulation and solar heat regulation. Additional green initiatives, such as rainwater harvesting, make CH2 a precedent-setting eco-friendly building, demonstrating how architecture can be inspired by nature to achieve sustainability (Fig. 2.61).

Fig. 2.61 CH2, Melbourne, Australia (2006), and sustainable façade applications

Location	Melbourne, Australia
Köppen Climate Classification	Oceanic climate (Köppen: Cfb)
Completion Date	2006
Building Owner/Primary Tenant	City of Melbourne
Architects & Design Teams	DesignInc
Number of Floors	10
Gross Floor Areas	12,500 m^2

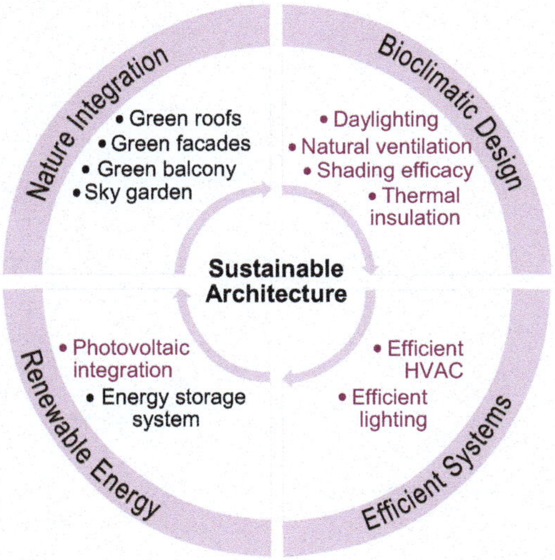

Chapter 3
Engineering Principles

Abstract This chapter discusses the fundamental engineering principles of sustainable tall building enclosures, emphasizing continuity as a design and engineering approach to achieve uninterrupted and coordinated performance across wall assemblies and all material and system interfaces. The continuity principles discussed in the chapter include structural load paths, fireproofing resistance, waterproofing protection, and thermal envelope performance. Continuity in structural load paths ensures that all structural loads are transferred uninterrupted through building enclosure components—from façade glazing and opaque panels to framing, then through façade brackets to anchor brackets embedded in the primary structure. Fireproofing continuity ensures the elimination of voids, gaps, or discontinuities within fire-resistive layers and associated fire and smoke stopping details. Fire- and smoke-safing wall assemblies, along with perimeter fire and smoke containment details, must function as a continuous protective system to prevent the fire and smoke spread between floors and compartments. Waterproofing continuity ensures uninterrupted water and vapor barriers across wall assemblies and interfaces, directing rain and condensation away from the building perimeter while resisting water and vapor intrusion. Continuity in thermal enclosures maintains continuous thermal insulation layers and minimizes thermal bridging to enhance energy efficiency and thermal comfort. Any breaks or discontinuities within the thermal layers result in condensation risks and mold growth, which negatively affect indoor air quality and occupant health. In summary, continuity in tall building enclosures ensures that wall assemblies function as uninterrupted, integrated systems for structural performance, life safety, waterproofing, and energy efficiency.

Keywords Tall building enclosures continuity · Structural performance · Fireproofing resistance and life safety protection · Waterproofing defense · Thermal envelope

Tall building enclosure technology and design have advanced significantly in recent years due to their multifaceted design and engineering requirements. With the latest innovations, building enclosures are now able to provide high performance

K. H. Kim, *Continuity in Sustainable Facades*,
https://doi.org/10.1007/978-3-032-19841-9_3

Table 3.1 Building enclosure performance requirements

Primary design goals of building enclosures	Building enclosure performance requirements
Protect	• Structural performance—strength and stiffness • Life safety—fire resistance • Impact resistance • OSHA requirements • Blast resistance • Water resistance • Condensation resistance • Sound protection—Outdoor Indoor Transmission Class (OITC)
Promote	• Energy efficiency—U-factor, R-value, Solar Heat Gain Coefficient (SHGC), Visible Light Transmittance (VLT), airtightness, natural ventilation, daylighting, bioclimatic assemblies • User experience—glare control, view-out
Re/generate	• Renewable energy integration • Nature integration • Carbon sequestration

and multiple functionalities, including structural integrity, life safety, high energy efficiency, enhanced safety and security, sustainability, and increased occupant comfort. In addition, new materials and design approaches enable more creative solutions for developing unique architectural features and contributing to contemporary cityscapes.

Building enclosures must satisfy a wide range of performance requirements, as outlined in Table 3.1. They must ensure adequate structural integrity under various external loading conditions while limiting deflection to a degree that is not perceptible to occupants. Protecting the public is the utmost priority and building enclosures must be fire resistance and prevent the transmission of fire and smoke. Additionally, building enclosures must also meet OSHA (Occupational Safety and Health Administration) requirements by providing safe working environments and protecting workers, installers, maintenance crews during fabrication, installation, operation and maintenance. Building enclosures also play an important role in promoting overall building energy efficiency and enhancing user experience by incorporating high performance and bioclimatic attributes, including good insulation, daylight provision, natural ventilation, and natural cooling and heating. To achieve carbon neutrality goals, building enclosures serve as a primary location for harnessing renewable energy and integrating nature due to their direct exposure to and interaction with the outdoor environment.

3.1 Continuity: Structural Load Path

Engineering of tall building enclosures requires a balance between different design criteria such as structural strength and stiffness, environmental resistance, airtightness, waterproofing, durability, user comfort, safety, and aesthetics. The structural integrity of a façade system protects occupants and the public. Any point of failure of façade elements can turn them into deadly projectiles. Façade failures highlight the severe risks associated with neglecting façade maintenance. Building owners, managers, and authorities must prioritize rigorous inspection and maintenance protocols, incorporating redundant safety systems, high-quality materials, and prompt repairs based on regular professional assessments. Proactive façade maintenance can significantly reduce the risk of tragic incidents and ensure community safety—as the consequences of neglect are far too severe to ignore.

A nine-story building located in the southern region of Kyushu, Japan, underwent partial collapse on April 17, 2016 due to failures of both the primary structure and the façade [224]. The region was affected by a series of strong earthquakes, followed by rainfall and severe weather that caused some parts of the façade structure to collapse. Strong winds further caused the collapse of the façades and imposed significant strains on the primary structure. Protective measures have been undertaken to avert the repetition of such events, including recommendations to strengthen the connection between the primary structure and building envelope systems. Such measures improve the safety of the building against harsh weather and seismic conditions.

Another case of structural failure occurred when an apartment building in Brooklyn experienced a partial structural collapse on November 4, 2022, leading to the evacuation of four residential units. This building was built in 1974, and exhibited a chronic façade failure due to structural weakness, with repair work scheduled for the near future [225]. To prevent such occurrences, façade systems could be constructed with higher strength materials or designed with greater safety factors in mind. In addition, regular inspection and maintenance strategies can prevent hidden structural damage and the weakening of connections [226].

In Turkey, a strong earthquake with a high magnitude of 7.8 on the Richter scale, struck the city of Diyarbakir on February 6, 2023, causing over 1300 deaths [227]. Reports indicated that the use of aluminum façade systems fastened with strong bracketry to the backup wall could avoid such catastrophic failures in the future by the use of appropriate material types and adequate structural engineering. Properly engineered metal façades can limit structural damage and reduce human casualties. The findings emphasized that catastrophic failures can be avoided through the use of appropriate façade materials and sound structural engineering practice. Aluminum façade systems with strong bracketry, properly fastened to the backup wall, can help avoid such failures. These strong façades are better able to withstand severe structural damage and reduce human casualties during a failure [228].

3.1.1 Design Criteria and Standards

ASCE 7 defines loads as "*forces or other actions that result from the weight of all building materials, occupants and their possessions, environmental effects, differential movement, and restrained dimensional changes*". Loads acting on building façades are mainly generated by various loading types, such as gravity load, lateral load, environmental load, impact load, and maintenance load. Gravity load can consist of dead load (e.g., self-weight of façade materials) and live load (e.g., occupant and furniture). Lateral loads mainly refer to wind load and seismic loads, where wind load acts perpendicular to the façade surface and seismic load acts in plane with the façade surface. Environmental loads include temperature, humidity, solar radiation, and barometric pressure due to the climatic conditions and site altitude. The structural loads discussed in the following sections are explained from the International Building Code and ASCE 7 [229].

Dead Loads: The dead load of façade materials is determined by unit material density or weight and material quantity and can be expressed as a point load (e.g., kg or lbs) or uniform load (e.g., kPa or psf). The dead load of a façade system consists of weight of façade materials including but not limited to metals, glass, silicone, mineral wool insulation, stone, dry gaskets and so on. The dead load of a façade system can fall on the order of 10–40 psf (0.48–1.92 kPa) depending on project conditions such as floor-to-floor height, material use, and geographic location. A margin of safety to the dead load of the façade system should be considered that account for miscellaneous components such as dry gasket, fasteners, sealants, and other accessories.

The weight of a façade system transfers to the primary structure through façade bracketry attached at the floor slab edge. The façade weight may cause deflection of slab edge or spandrel beam. When the slab edge deflection is expected to be too large, the spandrel beam may have a cambering process to counteract slab sagging under the self-weight of a façade system.

Live Loads: All gravity loads other than dead load are live loads. The live load of a façade system could be short-term or long term and a point load or uniform load. Examples include an occupant(s) standing on a horizontal façade transom, movable furniture, and equipment resting on façade frames. ASCE 7 provides live loading conditions for different occupancy and building uses. Live loads for a façade system should be determined based on professional judgement and project-specific conditions.

Wind Loads: Wind loading for façade design varies depending on wind speed, site location, building shape and size, façade unit size, and building program. When wind interacts with a tall building, positive pressure is applied on windward façades whereas leeward (suction) wind develops on the leeward façades and roof planes. Tall building enclosures should be designed for strength and serviceability to avoid wind-induced failures and to minimize deflection, inter-story drift and wind-induced vibration respectively.

The magnitude of the design wind load of a façade system depends on basic wind speed, exposure category of the building site, topographic effects of the site, building height, the effective wind area of the façade system, and internal pressure. ASCE 7 defines basic wind speed as 3-s gust wind measured at 33 ft. (10.1 m) above grade in Exposure C, such as an airport. Exposure category is classified as Exposures B, C, and D depending on ground surface roughness affecting air flows. Wind is generally stronger in open terrain than dense urban areas. Exposure B includes urban, suburban, and wooded areas with closely spaced single family houses or larger buildings. Exposure C is open terrain with scattered buildings less than 30 ft. (9.14 m) tall. Exposure D includes unobstructed flat terrain adjacent to water surfaces, mud flats, and unbroken ice.

Wind load varies with abrupt topography caused by isolated hills, ridges, and escarpments; therefore, the topographic effects of the building site should be considered when designing a façade system. Furthermore, wind load increases as the building height rises. The drag effect on the wind flow develops near the ground level due to the surface friction from the roughness of the ground surface. The drag effect gradually reduces as the building height is taller. Due to aerodynamic interaction between wind and buildings, the edge zone of the building (approximately 10% of the building width) is susceptible to higher wind load than the inner zone of the building. The building façade system at higher floor levels and edge zones is subject to higher wind loads compared to non-edge-zone façades at lower levels.

Wind acting on a building creates both positive and negative internal pressures. When the opening of windward façades is larger than the opening of the leeward façades, the interior space experiences positive internal pressure. On the other hand, when the opening area of the windward façades is smaller than that of the leeward façades, the interior space is subject to negative internal pressure. When the interior space is pressurized, all walls and roof are pulled inward. As a result, suction wind on the leeward side walls and roof combined with positive internal pressure produces higher net suction loads. Likewise, when the interior space is depressurized, all walls and roof will be pulled inward. Therefore, wind load on the windward side wall combined with negative internal pressure results in higher net wind load on that wall.

Other aerodynamic forces can create localized vortex shedding, resulting in wind-induced vibration and/or noise of a façade system, especially for external elements such as shading devices or catwalks. When the natural frequency of an external shading device, for example, matches the frequency of vortex shedding, resonance of the shading device occurs. The natural frequency of a façade system can differ depending on the mass and stiffness of the façade.

Flood Loads: ASCE 7 prescribes flood loads and flood damage considerations for building enclosures and structures. Under flood conditions, structures are influenced by both hydrostatic and hydrodynamic loads. The correct application of these design principles is necessary to protect buildings and building façades from flood-related catastrophes.

Hydrostatic loads are attributed to water levels up to the Design Flood Elevation (DFE) standard, defined as the maximum expected level of floodwater. All building surfaces that exist both above and below ground are required to resist hydrostatic loads. For surfaces that are constantly in contact with water, the DFE is increased by 1 ft. (0.30 m) to provide a safety margin for pressure increases due to prolonged exposure. Such a measure is important in preventing significant dangers posed by fluctuating water levels or sudden flooding.

Furthermore, hydrodynamic loads account for the dynamic effects of moving water, which must be carefully analyzed using fluid mechanics principles. To address these dynamic effects, an equivalent surcharge depth is added to the DFE design depth to accommodate the additional pressure generated by moving water. The resulting hydrostatic pressures are uniformly applied across the vertically projected area of the building or structure that is perpendicular to the flow. Additionally, surfaces parallel to the flow or in contact with slow-moving water (tail water) are also designed to resist hydrostatic pressures up to the DFE level.

The application of design principles enables engineers and architects to create flood-resistant structures which protect human life and reduce physical damage. Failure to adhere to these principles may lead to catastrophic consequences; therefore, knowledge and application of hydrostatic and hydrodynamic load factors must be fully integrated into building design to achieve robust and resilient built environments.

Seismic Loads: Earthquakes are catastrophic events that cause the death and destruction due to the sudden shaking of the earth. ASCE 7 explains seismic design requirements for building enclosures. The ground shaking generates a lateral force to oscillate a building back and forth in horizontal plane. Seismic resistant façade systems are designed to resist these lateral forces through controlled inelastic behaviors. The seismic design force is expressed as a percentage of weight of the façade system, which can range from a low percentage to as high as 50% of the façade weight. Earthquakes also generate vertical forces that may cause damage to the façade system. However, the self-weight of a façade system is typically greater than the vertical seismic forces, which rarely become governing seismic forces. Seismic design of building enclosures is technically challenging due to uncertainties and random variability, which sometimes require engineering judgement.

Rain Loads: ASCE 7 specifies rain loads for roof structures. Roof system design must account for water accumulation resulting from blocked or inadequate primary drainage systems. The roof is engineered to withstand the load from accumulated rainwater and the uniform load from water up to the level of secondary drain inlets. In areas prone to deep water accumulation, secondary (overflow) drains with separate lines may be preferable to overflow scuppers, thereby reducing the design rain load.

Free discharge is the preferred emergency drainage method. When calculating rain loads, it is assumed that the roof does not deflect, which simplifies the analysis but requires checking ponding instability separately. The hydraulic head, or depth of water above the secondary drainage inlet, depends on rainfall intensity, roof area serviced, and drainage system capacity.

In some regions, local regulations limit rainwater flow from roofs to storm drains. Controlled-flow drains are often used, requiring the roof to temporarily store stormwater. Roofs with controlled-flow drains are designed for a rain load of 30 lb./

ft^2 (1.44 kN/m^2) and include a secondary drainage system to prevent water depths exceeding 5.75 in (145 mm).

Environmental Loads: The façade system is the outermost building layer that experiences different temperature, air movement, relative humidity across a day and season. Environmental loadings such as temperature, air, and humidity (HAM—heat, air, and moisture) cause expansion and contraction of façade components. Every material has a specific coefficient of thermal expansion, which results in dimensional changes as material temperature varies. Façade components should be designed to accommodate dimensional movement freely under dynamic thermal conditions. As barometric pressure varies with job-site altitude, façade components—especially insulated glass units (IGUs)—may experience pressure imbalances between the IGU cavity and surrounding atmosphere. This can lead to failure of IGU hermetic seals, excessive oil-canning, and visual distortion.

Maintenance Loads: Building façades require periodic maintenance of façade cleaning and replacement. Façade maintenance systems typically consists of a building maintenance unit (BMU) on the rooftop, a hoisted BMU cradle, and fall protection or restraint systems. The suspended working cradle hoisted from the rooftop BMU is retrained by integrated curtainwall maintenance track. The façade therefore is subject to façade maintenance loads, which are assumed to act in multiple directions depending on maintenance activities. The design team must consider appropriate maintenance loads for the façade system design, considering the façade unit weight, required equipment, and the number of workers.

Impact Loads: Impact loads result from objects such as logs, ice floes, and debris striking buildings or structures. The U.S. Army Corps of Engineers categorizes into three levels: (1) normal impact loads from isolated, common objects; (2) special impact loads from large objects such as broken ice or debris; and (3) extreme impact loads from very large objects such as boats or collapsed buildings. Designing for extreme impact loads is generally impractical for most buildings. However, in high-risk locations, consideration of extreme impacts may be justified if the potential damage is severe. Designing for special and normal impact loads is more practical for most buildings and structures [230].

Standards and Building Codes: Tall building enclosures, like much of the construction industry, have become a global practice. Fabrication and construction techniques of façade systems have become increasingly standardized despite regional contextual and climate variations. Regional and national authorities have developed and adopted façade design and quality control standards to ensure the desired performance, user satisfaction, and design intent. Within the international community, including the United States, the European Unions, and China, notable industry standards have been adopted for façade engineering and executions.

In the United States, American Society of Civil Engineers (ASCE) and International Building Codes (IBC) regularly update structural design criteria for façade design. ASCE-7 *Minimum Design Loads and Associated Criteria for Buildings and Other Structures* serves the basis code for the structural design of façade systems in the United States. The American Society for Testing and Materials (ASTM) [231]

Table 3.2 Structural design standards and building codes for building enclosures

Standards and building codes
ASTM E330 Standard Test Method for Structural Performance of Exterior Windows, Doors, Skylights and Curtain Walls by Uniform Static Air Pressure Difference
ASTM E631 Terminology of Building Constructions
ASTM E997 Test Method for Evaluating Glass Breakage Probability Under the Influence of Uniform Static Loads by Proof Load Testing
ASTM E998 Test Method for Structural Performance of Architectural Glass Products Under the Influence of Uniform Static Loads
ASTM E1233/E1233M Test Method for Structural Performance of Exterior Windows, Doors, Skylights, and Curtainwalls by Cyclic Air Pressure Differential
ASTM E1300 Practice for Determining Load Resistance of Glass in Buildings
ASTM E1886 Test Method for Performance of Exterior Windows, Curtain Walls, Doors, and Impact Protective Systems Impacted by Missile(s) and Exposed to Cyclic Pressure Differentials
ASTM E1996 Specification for Performance of Exterior Windows, Curtain Walls, Doors, and Impact Protective Systems Impacted by Windborne Debris in Hurricanes
ASCE/SEI 7 Minimum Design Loads and Associated Criteria for Buildings and Other Structures

outlines testing protocols and material specifications for façade system and materials. ASTM E330 *Standard Test Method for Structural Performance of Exterior Windows, Doors, Skylights and Curtain Walls by Uniform Static Air Pressure Difference* outlines the evaluation of deflection and stress in façade assemblies under static wind pressure.

ASTM E330 mandates test specimens to be no less than two full building stories in height and include all pertinent framing, glazing, and structural anchoring components. Wind pressures are determined based on geographic location following International Building Code (IBC) wind maps and are imposed on test specimens at 2–10 s intervals, depending on the specimen's material properties and likelihood of fatigue [232]. Wind-loading protocols in the United States mandate all façade designs be capable of withstanding proof loading conditions, defined as loads 1.5 times greater than the design wind load for the location, and resist impact or airborne debris where required. While the upper portions of tall buildings must withstand higher wind-loading conditions, the lower portions of building façades are subject to more stringent impact-resistance requirements, particularly in hurricane- or tornado-prone areas. Table 3.2 summarizes structural design standards and building codes applicable to the engineering of building enclosures.

3.1.2 *Structural Performance Metrics*

Strength and Stiffness: Innovations in curtainwall have allowed architects to be more creative in the design of tall building façades. The curtainwall system is hung from the primary building structure. It does not carry gravity loads from floor, yet must remain strong and rigid enough to resist its own gravity load and lateral loads. Tall buildings sway and each floor is subject to story drift under lateral load. Each floor is subject to live load under lateral loading, and each floor is subject to interstory drift. Each floor is also subject to vertical movement under live loads. As a result of building movement, the façade system must remain isolated from the floor plates to avoid breakage, distortion, and disengagement from bracketry.

When designing façades for tall buildings, it is important to provide sufficient stiffness in exterior wall systems to maintain structural performance and ensure overall occupant comfort. Limiting the deflection of façade systems is important for several reasons. For instance, it prevents visible deformation that could undermine public confidence in a building's structural integrity. Excessive deformations increase the risk of material or system failure, potentially resulting in costly repairs and unsafe conditions. Considering these aspects collectively, adherence to stiffness requirements ensures that the external wall system performs structurally and functions efficiently while providing a weatherproofing barrier and preventing water and unwanted air infiltration. The International Building Code (IBC) sets strict deflection requirements for exterior walls to ensure the structural strength and durability. The IBC limits the deflection of exterior walls incorporating brittle materials to be between L/240 and L/360 under wind or snow loads, where L represents the span length of wall components [233].

Different occupants have various motion tolerances, and it becomes challenging to design a façade system that can satisfy everyone. H/360, where H represents floor-to-floor height, is commonly accepted as an allowable interstory drift that which occupants are unlikely to feel discomfort. A deflection limit of L/175 for a façade mullion length of 14 ft.–6 in is commonly recommended of façade system deflection.

Engineering of tall building enclosures should be understood in relation to the different structural loading conditions imposed on building enclosures. For example, curtainwall construction follows inherent façade design principles in which glazing units are supported by metal frames such as mullions and transoms, and the entire curtainwall assembly is anchored to the primary building structure, typically at slab edges or spandrel beams. The interface between glazing units and metal frames within the curtainwall unit may be mechanical, structural silicon-based, or a hybrid of both connection types. The curtainwall bracket is typically preinstalled at the top of the mullion. Its connection to the primary structure may occur at the top, face, or underside of the slab, depending on curtainwall set out and structural requirements.

Figure 3.1 explains how curtainwall components resist structural loads and maintain structural integrity under various structural loading conditions. The glazing system initially receives the structural loads, which are transferred to the horizontal framing members and subsequently to the vertical framing members. The curtainwall bracket, preinstalled at the top of the vertical mullions, is then connected to anchor brackets embedded in primary structure such as slabs or spandrel beams.

This attachment method renders the curtainwall unit as a thin glass skin capable of resisting self-weight and lateral loads while effectively transferring structural reactions to the primary structure. The structural design of the tall building enclosures is primarily determined by gravity and lateral loads. Gravity loads include dead load (self-weight), maintenance loads, and live loads associated with occupants, interior furniture, and maintenance equipment. Lateral loads primarily include wind and

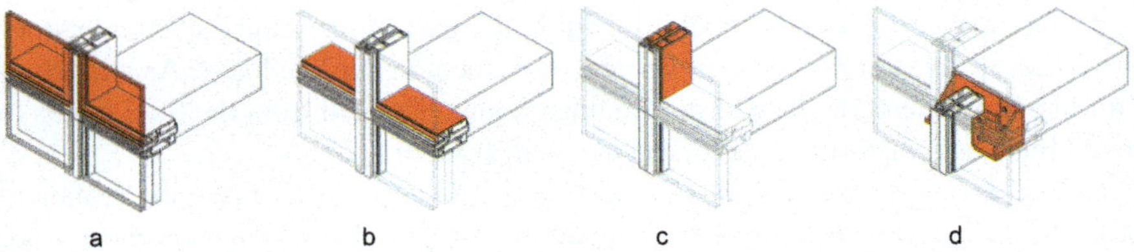

a. The glazing unit transfers the gravity load (e.g., dead load).
b. The gravity load (e.g., dead load) is distributed from the glazing unit along the horizontal frame (transom).
c. The gravity load (e.g., dead load) carried by the horizontal frame is transferred to the vertical frame (mullion)
d. The pre-installed curtainwall bracket on the mullion transfers the gravity load (e.g., dead load) to the anchor brackets, which subsequently transfer the load to the slab.

Fig. 3.1 Continuous gravity load transfer from the curtainwall components to the anchor bracket

seismic loads acting on the curtainwall assembly. As far as structural reactions are concerned, structural reactions act in-out of façade unit and in-plane of façade unit respectively. For example, wind suction pulls the glazing unit away from the frame and imposes greater stress on frame connections and façade brackets. Seismic in-plane loading causes the in-plane deformation of the frame, straining the connection between the glazing unit and the framing components. Figure 3.1 illustrates the load path sequence for the curtainwall components under the gravity loads such as dead load.

The structural load path of gravity loads (e.g., dead load) is transferred continuously from the glazing unit to the horizontal frame, then to the vertical frame, then to the curtainwall bracket, and finally to the anchor bracket installed on the primary structure. For lateral loads such as wind pressure and suction loads, the glazing unit transfers the wind loads to the frame based on the tributary load area, which then transfers the loads to the curtainwall bracket and anchor brackets. Figure 3.2 highlights the structural load transfer path through essential curtainwall components, starting with the glazing unit, horizontal transoms, vertical mullions shows its sight profile, curtainwall brackets, and finally the anchor brackets.

Differential Movements: Differential movements in building systems and components play a crucial role in façade design and construction. Differential movements in façade design refer to the multi-directional movements between the façade system and primary structure under design loads such as wind, earthquake, and live load. These movements, caused by various loading conditions, necessitate careful consideration of the interaction between the façade system and the primary structure. The implementation of vertical and horizontal joints in split curtainwall extrusions provides essential flexibility to accommodate these movements. The use of adjustable façade brackets offers a practical solution for managing construction tolerances and installation variations. By incorporating these design elements, architects and engineers can ensure that building façades remain stable and functional under diverse structural loadings and environmental conditions, ultimately contributing to the overall safety and longevity of the façade system.

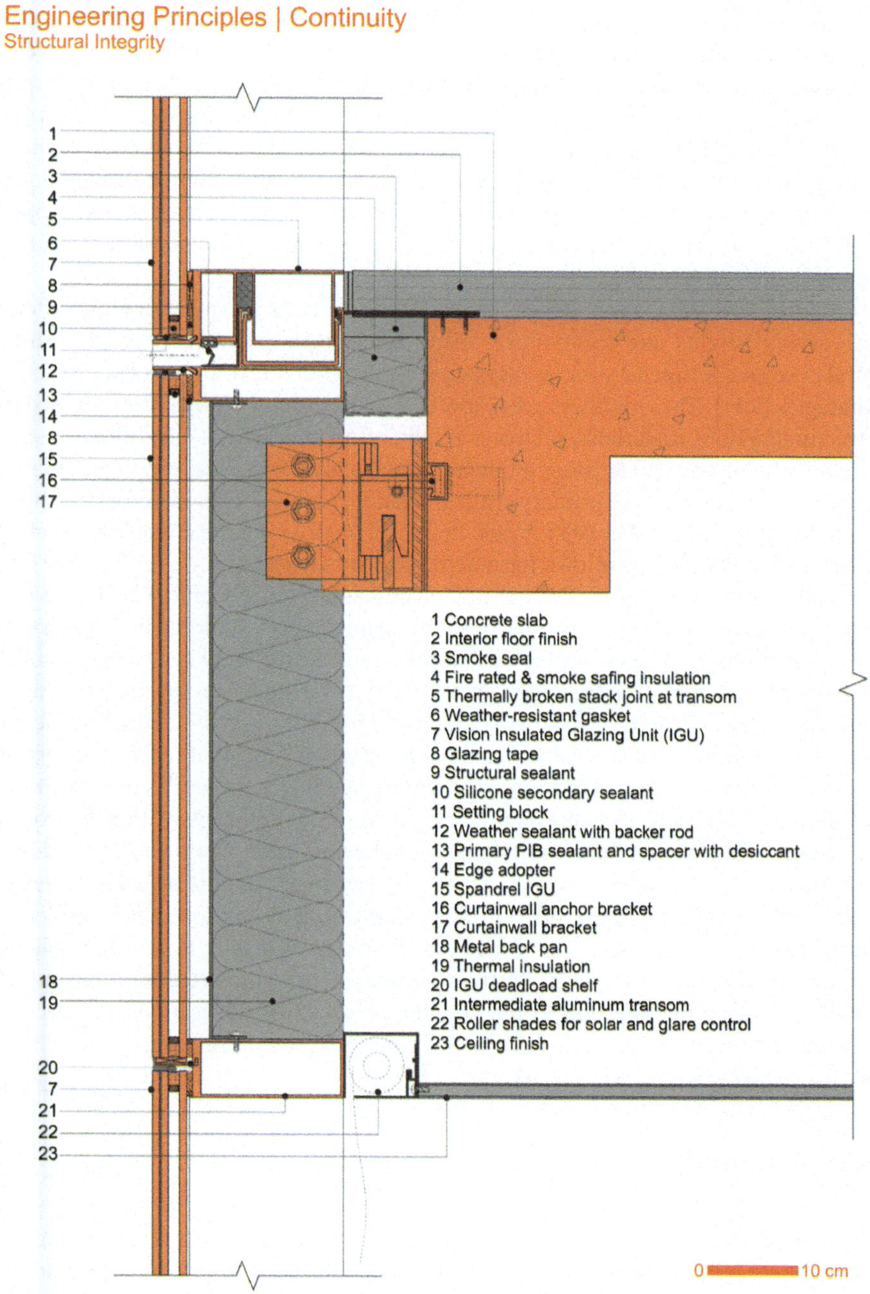

Fig. 3.2 Curtainwall components transferring various structural loads

Column Shortening: The design and construction of tall tower building enclosures should account for column shortening as it poses a significant challenge to the long-term serviceability of façade systems. Column shortening occurs primarily due to elastic shortening, shrinkage, and creep of reinforced concrete columns and foundations in tall towers, where elastic shortening is an instantaneous column shortening caused by material properties and the duration and magnitude of axial loads from floors above. Various conditions affect this process, including material properties, structural loads, construction sequence, and environmental conditions such as temperature and humidity.

Studies analyzing creep and shrinkage effects on tall building structural systems of varying heights aim to minimize differential displacement between the structural core and perimeter structural systems through holistic design and construction strategies [234]. These studies reveal how creep and shrinkage impacts vary based on structural system, materials, and tower height, offering practical, cost-effective solutions. Column shortening poses a threat to the integrity of a façade system. Holistic coordination between the primary structure and façade system is therefore essential. Appropriate joint size between façade units and flexible interface materials such as sealants and dry gaskets allows for independent movement.

Building Sway and Story Drift: A tall tower is subject to dynamic wind forces as it rises in height. Towers must respond to aerodynamic effects. Building façades receive pressure from windward and leeward winds, causing tower sway. Building shape and orientation as well as façade geometry and materials play a role in responding to building sway and story drift. Curvilinear forms are more aerodynamically efficient and stable in high winds, whereas square or rectangular forms with reentrant corners increase vortex shedding. The process of vortex shedding at a tower's corner imposes vibrations on building systems and increases overall building sway. Besides vortex shedding, wind creates other aerodynamic behaviors such as downdrafts, eddies, crosswinds, and quartering winds. These aerodynamic patterns affect both occupants and pedestrians. Story drift of the building should be properly controlled to mitigate excessive deformation of façade systems. In principle, façade components are designed to accommodate construction tolerances, which represent deviations of the primary structure due to the cumulative effect of fabrication and erection tolerances from its nominal position at the time of façade installation.

3.2 Continuity: Fireproofing Assemblies

Ensuring the fireproofing of building enclosures is essential for public safety. Implementing effective measures saves lives, protects property, and minimizes economic losses caused by fires. Utilizing fire-resistant materials and assemblies helps to prevent rapid fire and smoke spread, thereby preserving structural integrity even under extreme heat conditions. By establishing robust fireproofing strategies, buildings enable safe evacuation for occupants and facilitate more effective firefighting operations, thereby mitigating the overall impact of fire disasters.

The 2017 Grenfell Tower fire in London spread from the fourth floor to the top of the 24-story building and killed 72 people [235]. The fire spread rapidly due to aluminum composite material (ACM) rainscreen panels with polyethylene cores, which fueled the flames. Melting and dripping burning polyethylene from crown, spandrel, and column panels ignited fires lower levels, which then travelled upward, creating diagonal flame progression across the tower's façades. Polyisocyanurate (PIR) and phenolic foam insulation boards behind ACM panels, and possibly around window surrounds, further contributed to vertical flame spread. Another notable fire incident occurred at the Monte Carlo casino in Las Vegas in 2008 and involved EIFS wall panels with an inadequate acrylic finish thickness, as well as combustible decorative expanded polystyrene three-dimensional "shapes." The fire was most likely triggered by a welder's torch and spread upward and laterally across the façade of the upper floors of the building. Subsequent analysis showed that while the non-fire-rated expanded polystyrene decorations served as the primary fuel load, the EIFS also contributed significantly to the fire. As with any other fire-rated assembly, it is important that EIFS be fabricated and installed exactly as tested to ensure the intended fire-resistance performance.

3.2.1 Design Criteria and Standards

Walls need to resist gravity loads and lateral forces. Gravity loads transfer compressive forces from the top or from various points along the wall to the ground. Wind loads cause out-of-plane bending forces on walls.

Walls can be categorized into load-bearing or non-load-bearing walls. Load-bearing walls carry the weight of part of the structure. For example, the wall of the Pantheon carries the roof load. Non-load-bearing walls support only their own weight. Examples of non-load-bearing walls include contemporary enclosures such as curtainwalls, veneer walls, and panel walls. The collapse of a load-bearing wall is more serious than that of a non-load-bearing wall because it supports other building elements. However, for a firefighter caught in a collapse, it makes little difference whether the wall is load-bearing or not.

Fire weakens or destroys structural elements, overloading the structure and potentially causing unpredictable collapse. Some building systems require a certain duration of fire resistance and must be able to withstand fire exposure without collapsing. Many building components are non-fire-resistant and may easily collapse when exposed to a major fire. Combustible materials have different caloric values (Table 3.3). Common caloric value estimates are 24 MJ/kg for wood and paper, and 26 MJ/kg for polyurethane foam insulation.

Fire Codes: Building regulations have evolved incrementally, often in response to specific fire disasters. In the nineteenth century, cities enacted laws to address issues

Table 3.3 Caloric values of building enclosure materials (reproduced from Bachanan [236])

Materials	Combustion heat (MJ/kg)
Woods	17–20
Particle boards	17–18
Straws	15–16
Cardboard	13–21
Rubber isoprene	44–45
Plastics	
ABC	34–40
Acrylic	27–29
Celluloid	17–20
Epoxy	33–34
Polyester	30–31
Polyethylene	43–44
Polystyrene	39–40
Polyisocyanurate foam	22–26
Polycarbonate	28–30
Polypropylene	42–43
Polyurethane	22–24
Polyurethane foam	23–28

such as adequate building exits (e.g., exterior fire exit stairs), the need for fire walls, and restrictions on wood-frame construction in certain areas.

In the 1870s, Boston Fire Chief John Damrell advocated national building regulations following the Great Boston Fire of 1872. However, it was not until the Great Baltimore Fire of 1904 that a nationwide effort emerged to develop a model building code. Since then, the United States has built upon these early efforts to establish the current national model building codes.

Codes have primarily focused on general fire safety for the community and occupant life safety during evacuation. Fire codes regulate activities in existing buildings, including the maintenance of fire protection features, hazardous processes, and the storage of hazardous materials. Additionally, they require life safety systems such as sprinklers, fire walls, fire exit doors, and fire stairs. However, building codes are usually not retroactive; therefore, existing buildings may be nonconforming and potentially hazardous.

There are five main types of construction regulated in the International Building Codes (IBC), and each construction type has allowable building height and floor area limits. In addition, building elements are assigned fire-resistance ratings in accordance with ASTM E119 and UL 263. IBC requires different fire-resistance rantings for building elements depending on construction types. Type I and Type II construction utilize varying fire-resistance ratings for exterior walls as specified in IBC's Table 601, except as permitted in IBC's Table 603. Type III construction requires exterior walls to have a fire-resistance rating of two hours as specified in IBC's Table 601. Type IV construction involves mass timber construction, with fire-resistance ratings per IBC's Table 601 [237].

Flow paths, defined as the movement of hot fire gases and smoke from high-to low-pressure areas, can occur when windows or doors are opened or fail. Providing these flow paths allows oxygen to enter the fire compartment, thereby fueling fire growth. A fire-resistance rating of building systems is determined in accordance with ASTM E119 *Standard Test Methods for Fire Tests of Building Construction and Materials or UL 263 Fire Tests of Building Construction and Materials.* Additionally, noncombustible materials are determined in accordance with ASTM E2652 *Standard Test Method for Assessing Combustibility of Materials Using a Tube Furnace with a Cone-shaped Airflow Stabilizer*, at 750 °C. Materials with surface thickness less than 1/8 in (3 mm) and having a flame spread index 50 or less, when tested in accordance with ASTM E84 or UL, 723 shall be considered noncombustible. Fire resistance in building materials is crucial for protecting occupant safety by slowing fire spread and allowing safer evacuation. It provides structural integrity and helps prevent building collapse during fires while limiting fire damage and associated financial loss. Table 3.4 explains different fire-resistance characteristics for building enclosure materials.

Building codes ensure fire-resistance ratings of building materials and construction assemblies and establish the required fire separation distances between buildings. Fire-resistance ratings are designed to withstand fire exposure inside buildings, ensuring the safety of occupants during evacuation. Fire separation distances between neighboring exterior walls are intended to reduce the risk of fires spreading to adjacent buildings. These fire-resistance measures also safeguard firefighters by enabling safer emergency response operations while minimizing their exposure to fire hazards. Table 3.5 summarizes fire separation distance requirements for exterior walls based on construction types and occupancy types.

The fire codes and industrial standards ensure that building materials and construction assemblies will withstand and function effectively under fire-emergency conditions. Industrial standards developed by the American Society for Testing and Materials (ASTM) are widely referenced in regulating building codes. ASTM E119 *Standard Test Methods for Fire Tests of Building Construction and Materials* is a widely accepted test used to classify the hourly fire-resistance ratings of load-bearing architectural elements such as load-bearing walls, columns, beams, and floors. Furthermore, ASTM E84 *Standard Test Method for Surface Burning Characteristics of Building Materials* has become an essential guideline for the evaluation of flame-spread properties of numerous building materials (Table 3.6).

3.2.2 Fire-Resistance Performance Metrics

Combustibility and smoke toxicity are important fire-resistance metrics. Following the Grenfell Tower fire tragedy, where combustible façade materials allowed rapid fire

Table 3.4 Fire-resistance characteristics of building enclosure materials (reproduced from Brannigan [238])

Material types	Material behaviors under fire
Metal	• Aluminum melts at typical fire temperatures and loses strength rapidly when heated • Galvanized cold-formed steel members are noncombustible but may lose strength and fail when exposed to significant building fires
Bricks	• Bricks are fire-resistant masonry units made of cured clay. However, they may spall when exposed to fire, losing structural integrity and potentially leading to collapse
Natural stones	• Natural stones such as granite, marble, limestone, and sandstone are used for walls and foundations. These materials can spall (fracture or break off in in pieces) when exposed to fire, with granite being particularly susceptible
Concrete	• Concrete generally has good fire resistance but may lose compressive strength under fire
Wood	• Heavy timber can develop a char layer to maintain structural integrity when it burns • Fire-retardant salts can be pressure-impregnated into wood to improve fire resistance of wood panels
Glass	• Glass has little resistance to heat or pressure buildup from a fire. Windows often fail violently during severe fires, sometimes appearing as explosive-like breakage • Some windows are double- or triple-glazed for energy efficiency, making them difficult for firefighters to break when emergency access through windows is required. Some building codes require designated windows in high-rises to be breakable for emergency access, marked with a fire-rescue symbol
Composites	• Aluminum-clad polyethylene panels are popular for building enclosure finishes. Their shiny metal surfaces conceal a core of flammable polyethylene that can liquefy and ignite when exposed to heat • Gypsum wallboard is a common building material used in firewalls and partitions. It consists of calcium sulfate dihydrate pressed between cardboard sheets. Wallboard assemblies must be tested as a system to get a fire-resistance rating • Gypsum board may contain additives such as plastic materials or fiberglass mesh to alter its properties. Breaching these materials can be difficult, but the connection points at sheet ends are weak spots and may be exploited. Cutting through the board requires caution due to the entanglement hazard posed by fiberglass mesh
Insulation	• Thermoplastics can produce flaming, dripping molten plastic, even in fire-rated assemblies, whereas thermosets char and burn without flowing • Glass fiber insulation contains non-combustible glass fibers and combustible binders. It is often installed inside the curtainwall spandrel panels and above suspended ceilings. If the insulation retains heat within the system assembly, the accumulated heat may cause the system to fail sooner than expected. It is important to use fire-rated glass fiber assemblies

(continued)

Table 3.4 (continued)

Material types	Material behaviors under fire
Exterior insulation finishes system	• EIFS, a multilayer exterior wall system, is composed of an inner substrate, an insulation board, a water-resistant barrier, fiberglass mesh, a base coat, and cladding finish • The EIFS'S finished surface is adhered to the insulation board and water-resistance layer • Because a non-fire rated assembly can act as a significant fuel source, the EIFS must adhere to applicable fire-safety guidelines during both fabrication and installation

Table 3.5 Fire-resistance rating requirements for exterior walls based on fire separation distances (IBC Table 705.5) [239][a,d,g]

Fire separation distance = X (feet)	Type of construction	Occupancy Group H[e]	Occupancy Group F-1, M, S-1[f]	Occupancy Group A, B, E, F-2, I, R[i], S-2, U[h]
X < 5[b]	All	3	2	1
5 ≤ X < 10	IA, IVA	3	2	1
	Others	2	1	1
10 ≤ X < 30	IA, IB, IVA, IVB	2	1	1[c]
	IIB, VB	1	0	0
	Others	1	1	1[c]
X ≥ 30	All	0	0	0

For SI units: 1 ft. = 304.8 mm

[a] Load-bearing exterior walls shall also comply with the fire-resistance requirements of Table 601

[b] See Section 706.1.1 for party wall requirements

[c] Open parking garages that comply with Section 406 shall not be required to have a fire-resistance rating

[d] The fire-resistance rating of an exterior wall is determined based on the fire separation distance of the exterior wall and the story in which the wall is located

[e] For special requirements for Group H occupancies, see Section 415.6

[f] For special requirements for Group S aircraft hangars, see Section 412.3.1

[g] Where Table 705.8 permits nonbearing exterior walls with unlimited areas of unprotected opening, the required fire-resistance rating for the exterior walls shall be 0 h

[h] For a building containing only Group U occupancy, including private groups or carports, the exterior wall shall not be required to have a fire-resistance rating where the fire separation distance is 5 ft. (1524 mm) or greater

[i] For a Group R-3 building of Type II-B or Type VB construction, the exterior wall shall not be required to have a fire-resistance rating where the fire separation distance is 5 ft. (1524 mm) or greater

Table 3.6 ASTM testing for fire-resistant materials and assemblies [240, 241]

Test number and test type
ASTM A390 Specification for Zinc-Coated (Galvanized) Steel Poultry Fence Fabric (Hexagonal and Straight Line)
ASTM C1186 Specification for Flat Fiber-Cement Sheets
ASTM C1288 Specification for Fiber-Cement Interior Substrate Sheets
ASTM C1396/C1396M Specification for Gypsum Board
ASTM C6513 Practice for Calculating the Superimposed Load on Wood-frame Walls for Standard Fire-Resistance Tests
ASTM D4442 Test Methods for Direct Moisture Content Measurement of Wood and Wood-Based Materials
ASTM D4444 Test Method for Laboratory Standardization and Calibration of Hand-Held Moisture Meters
ASTM E69 Test Method for Combustible Properties of Treated Wood by the Fire-Tube Apparatus
ASTM E119 Standard Test Methods for Fire Tests of Building Construction and Materials
ASTM E160 Test Method for Combustible Properties of Treated Wood by the Crib Test 3
ASTM E162 Test Method for Surface Flammability of Materials Using a Radiant Heat Energy Source
ASTM E176 Terminology of Fire Standards
ASTM E177 Practice for Use of the Terms Precision and Bias in ASTM Test Methods
ASTM E691 Practice for Conducting an Interlaboratory Study to Determine the Precision of a Test Method
ASTM E814 Test Method for Fire Tests of Penetration Firestop Systems
ASTM E2032 Practice for Extension of Data From Fire Resistance Tests Conducted in Accordance with ASTM E119
ASTM E2231 Practice for Specimen Preparation and Mounting of Pipe and Duct Insulation Materials to Assess Surface Burning Characteristics
ASTM E2404 Practice for Specimen Preparation and Mounting of Textile, Paper or Polymeric (Including Vinyl) and Wood Wall or Ceiling Coverings, Facings and Veneers, to Assess Surface Burning Characteristics
ASTM E2573 Practice for Specimen Preparation and Mounting of Site-Fabricated Stretch Systems to Assess Surface Burning Characteristics
ASTM E2579 Practice for Specimen Preparation and Mounting of Wood Products to Assess Surface Burning Characteristics
ASTM E2599 Practice for Specimen Preparation and Mounting of Reflective Insulation, Radiant Barrier and Vinyl Stretch Ceiling Materials for Building Applications to Assess Surface Burning Characteristics
ASTM E2652 Standard Test Method for Assessing Combustibility of Materials Using a Tube Furnace with a Cone-shaped Airflow Stabilizer, at 750 °C
ASTM E2688 Practice for Specimen Preparation and Mounting of Tapes to Assess Surface Burning Characteristics
ASTM E2690 Practice for Specimen Preparation and Mounting of Caulks and Sealants to Assess Surface Burning Characteristics
ASTM E2768 Test Method for Extended Duration Surface Burning Characteristics of Building Materials (30 min Tunnel Test)
ASTM E2988 Practice for Specimen Preparation and Mounting of Flexible Fibrous Glass Insulation for Metal Buildings to Assess Surface Burning Characteristics
ASTM E3202 Practice for Specimen Preparation and Mounting of Plastic Composites for Use as Deck Boards, Stair Treads, Guards or Handrails to Assess Surface Burning Characteristics
ASTM E3287 Practice for Specimen Preparation of Fenestration Profiles Intended to Support Non-Combustible In-Fill Materials to Assess Surface Burning Characteristics
NFPA 262 Standard Method of Test for Flame Travel and Smoke of Wires and Cables for Use in Air-Handling Spaces (2019)
NFPA 285 Standard Fire Test Method for Evaluation of Fire Propagation Characteristics of Exterior Wall Assemblies Containing Combustible Components
UL 263 Fire Tests of Building Construction and Materials
UL 723 Standard for Test for Surface Burning Characteristics of Building Materials (2023)

to spread, killing 72 people, the UK banned combustible materials on tall building exteriors [242]. Plastic insulation has increased smoke toxicity compared to non-combustible products. Additionally, the ignition of polyisocyanurate (PIR) foam insulation by flaming polyethylene (PE) drips from aluminum composite materials highlights that façade systems cannot be treated as simple sum of their individual components, as components can interact synergistically. Polyisocyanurate insulation also produces smoke toxicity approximately 15 times greater than mineral wool. New buildings increasingly require fire safety measures, while energy-efficiency goals drive the use of flammable insulation materials. Most fire-related fatalities result from inhalation of toxic gases. A study examined the fire toxicity of six insulation materials under various conditions using Fractional Effective Dose (FED) and Lethal Concentration 50% (LC50) models [243]. Plastic insulations, including polyisocyanurate and polyurethane foam, and hydrogen cyanide, were found to release elevated levels of toxic gases, significantly increasing toxicity under unventilated conditions. The materials showed increasing toxicity ranging from stone wool (least toxic) to polyisocyanurate foam (most toxic). Another study discussed façade flame spread hazards in high-rise buildings due to three primary mechanisms: fire leapfrogging via window breakage, failure of fire stopping between floors and exterior walls, and radiant heat flux impacts on façade assemblies [244]. To mitigate this hazard, it was recommended to determine the enclosure fire size, ejected façade flame properties (height and heat flux), and their impacts on façade assemblies.

Fire stops are another fire-resistance measure. Modern buildings frequently have poor perimeter firestopping and numerous penetrations for wiring and air conditioning, which can serve as smoke paths. Concrete floors with holes can allow deadly fire gases to pass through. Building floors often require penetration for utilities, plumbing, HVAC, electrical, and communications. Proper methods, like firestopping or fire-rated shafts, should be used to maintain floor integrity. Concrete floors require expansion joints, which can create fire transmission risks if made of steel, aluminum, or combustible materials. Concrete shrinkage can also lead to large cracks, allowing fire to spread between floors, even if the concrete is structurally sound. Shrinkage, distortion, or settlement may also misalign door frames, preventing fire barrier doors from closing properly.

The curtainwall components are required to provide fire protection and smoke control to protect public life safety and comply with building regulations. They consist primarily of glass, metal, and mineral wool insulation, all of which are considered noncombustible and exhibit high-temperature resistance. The space between the rear surface of the spandrel panel and the face of the floor slab is sealed with fire-rated gaskets and sealants. These firestopping details are intended to restrict the spread of both the fire and smoke between the floors. Figure 3.3 illustrates curtainwall components that provide fire resistance and smoke control.

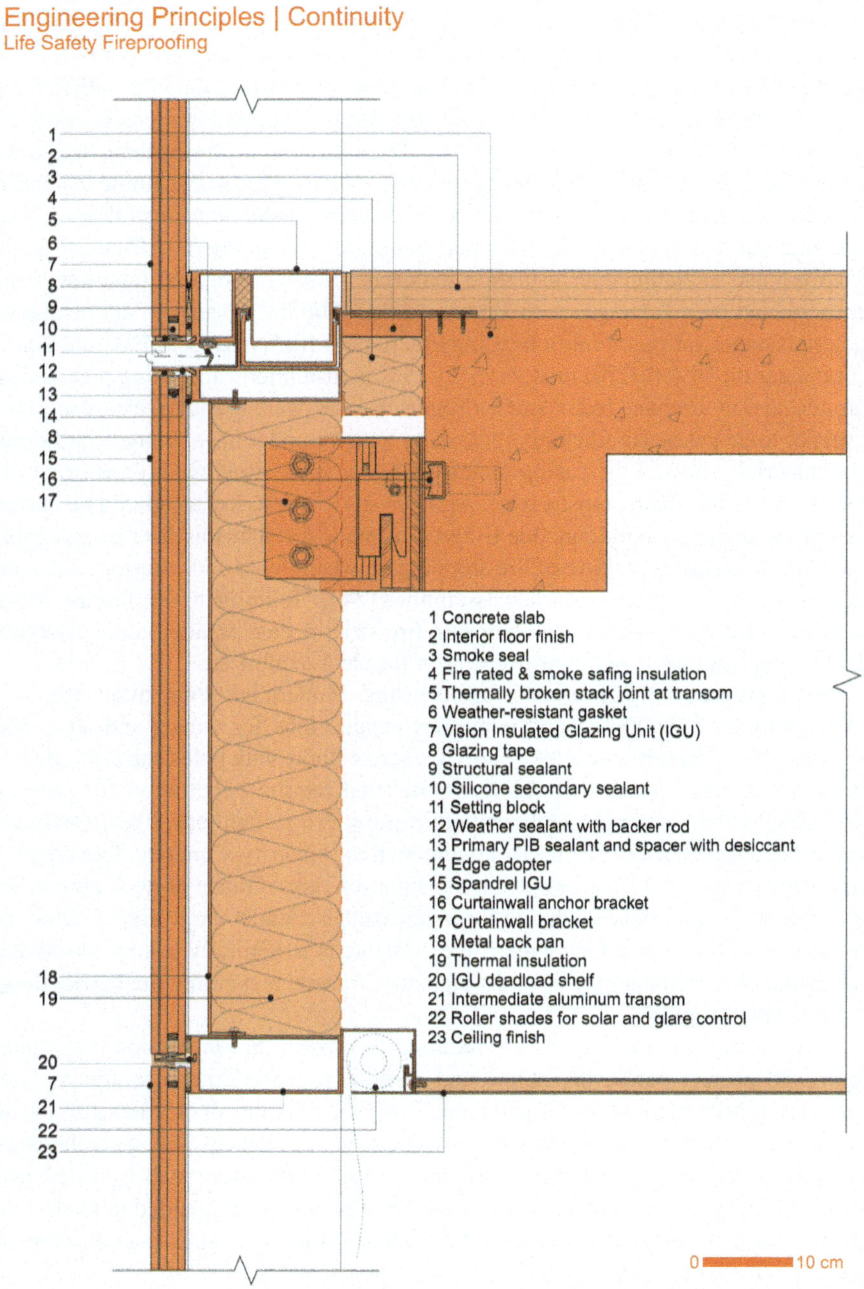

Fig. 3.3 Curtainwall components with fire resistance and smoke containment details

3.3 Continuity: Waterproofing Assemblies

Tall building enclosures are an important element of a building's overall performance, safety, and aesthetic appearance. They provide structural integrity, serve as water barriers, improve energy efficiency, reduce noise levels, offer additional security, and increase aesthetic appeal. Tall tower enclosures need to provide structural integrity under various external and internal loads. The material selection and assembly method are chosen to provide strength and stiffness under extreme load cases, allowing architects and engineers to build taller and aesthetically articulated structures. Due to site constraints and accelerated construction schedules, tall enclosures are typically configured using a prefabricated construction approach made of metal frames infilled with panels of various materials such as glass, metal, or other lightweight materials. By using prefabricated construction techniques, tall towers can be built more quickly with improved quality control. Prefabricated curtainwall assemblies must be able to withstand the forces from their own self-weight, lateral loads such as wind and earthquakes, environmental loads, and other extreme weather conditions and natural disasters. Tall building enclosures require architectural design and engineering considerations to ensure that their structural capability resists all applicable loads while also providing aesthetic appeal. They are the outermost exterior layers that protect buildings from wind and strong gusts and reduce environmental impacts on buildings. Metal frames infilled with glass, for example, provide a physical barrier between indoor and external environments while also offering a sleek architectural expression. There are a few factors that must be considered when selecting building enclosure typology and materials. Extreme structural loadings and climate conditions should be considered when selecting enclosure systems and materials. The aesthetic appeal of the enclosure is another important factor, as it is visible from multiple viewpoints across the site and surrounding city context.

3.3.1 Causes of Water Penetration

Buildings still face issues with water leakages despite technological advancement. This is due to the increasing complexity of construction, the disregard for basic waterproofing principles, and a lack of coordination between construction systems and materials. Controlling groundwater, rainwater, and surface water can prevent damage and costly repairs to building envelopes. Water is the most destructive weathering element for concrete, masonry, and natural stone structures, causing significant damage accompanied by structural degradation and mold formation, leading to mold-related health issues. Waterproofing techniques that account for natural forces and their effects can preserve a structure's integrity and health indoor environment. Proper water proofing details and material selection are crucial to counter the detrimental effects of these natural forces.

Water sources that can penetrate building envelopes typically include rainwater on above-grade components or groundwater intrusion at below-grade. Other sources, such as melting snow, cooling tower overspray, landscaping sprinklers, and redirected water from downspouts and gutters should also be considered. Water leakage occurs when three conditions must be met: (1) water presence, (2) a breach, such as holes or discontinuity, and (3) water force that moves water inward. Water is a critical factor in building enclosure performance. Small openings can provide pathways for water intrusion, influencing flow rate and direction. Surface tension, gravity, and capillary actions draw water into penetration, leading to structural damage and promoting mold growth.

3.3.2 Waterproofing Design and Construction Practices

Waterproofing practice begins with controlling potential sources of water leakage by actively directing groundwater away from the site, channeling rainwater away from the building, and diverting surface runoff from the building envelope. Most building materials and assemblies are vulnerable to water damage, including concrete, stone, masonry, metal, and other construction materials. During construction, it is important to implement prefabrication construction methods whenever possible to improve quality control. In addition, interfaces between dissimilar systems and materials should be carefully detailed and joined using advanced waterproofing membranes, high-performance sealants and gaskets, and improved drainage and flashing systems. Small openings and holes from inadequate waterproofing construction can become routes for water intrusion which can subsequently lead to property damage and water-induced mold contamination with building assembles and interior surfaces.

Careful attention must be paid to the design and detailing of the building envelope to mitigate the risks posed by water (Table 3.7). Proper waterproofing details such as metal flashing, membrane, sealant, and drip edge details are essential to create effective waterproofing assemblies. Water barrier, drainage, and diversion systems are essential for redirecting water away from vulnerable areas and preventing it from entering the building. Leaks are more often attributed to poor workmanship and improper field practices, such as insufficient coverage, inadequate substrate preparation, or application over uncured concrete, as well as a lack of attention to proper detailing at material terminations and transitions [245].

Table 3.7 Waterproofing techniques for building enclosures

Waterproofing materials	Waterproofing details	Waterproofing system
Metal	Sloped surfaces (avoid ponding)	Water barrier system
Polymer membranes	Cap flashing and counterflashing	Water drainage system
Sealants	Drip edges	Water diversion system

Curtainwall façades provide multiple functions. Among their key performance requirements, water- and airtightness are critical for preventing water leaks, aging materials, and user discomfort, as well as improving air quality, and energy efficiency. The key curtainwall components are glass, sealant, and dry gasket located within the air gap between the split framing members (Fig. 3.4). Thorough waterproofing details and high-quality workmanship are crucial to counter the detrimental effects of water leaks.

3.3.3 Design Criteria and Standards

ASTM was established in 1898 to set standards for materials, including waterproofing products like sealants and caulking. These standards allow comparison between different products and manufacturers. ASTM standards provide detailed test requirements to ensure materials are tested under uniform and controlled conditions. This enables reliable comparison of characteristics such as weathering resistance, performance, and installation conditions. ASTM develops standardized test methods and specifications for evaluating and comparing products (Table 3.8). Manufacturers document materials that have been tested in accordance with ASTM standards by independent laboratories under controlled conditions. These tests assess a material's performance capabilities and limitations, as well as its suitability for exposure to extreme environmental conditions [246]. When window assemblies are tested under ASTM E331, for example, no water leakage is permitted for 15 min when the window is subjected to a flow rate of 0.05 L/s m^2 with a differential pressure across the window of 298 Pa.

3.4 Continuity: Thermal Envelope

Tall building enclosures are an important part of the overall building performance, providing greater structural integrity and safety for tall buildings. They not only help to protect buildings from weather and other external environments but also contribute to comfortable and healthy indoor environments while reducing building energy consumption. Installing high-performance enclosures is becoming an increasingly popular strategy to increase building energy efficiency, lower energy costs, reduce environmental impacts, and support high levels of occupant satisfactions. Tall tower enclosures can take advantage of natural light and ventilation (where permitted) while still controlling indoor temperature, solar gain, and clean air supply within the building. This not only reduces energy bills but also provides a healthier and more productive work environment for occupants. Additionally, these enclosures can be customized to meet specific needs such as user-controlled shading devices, making them an effective design approach for enhancing occupant productivity.

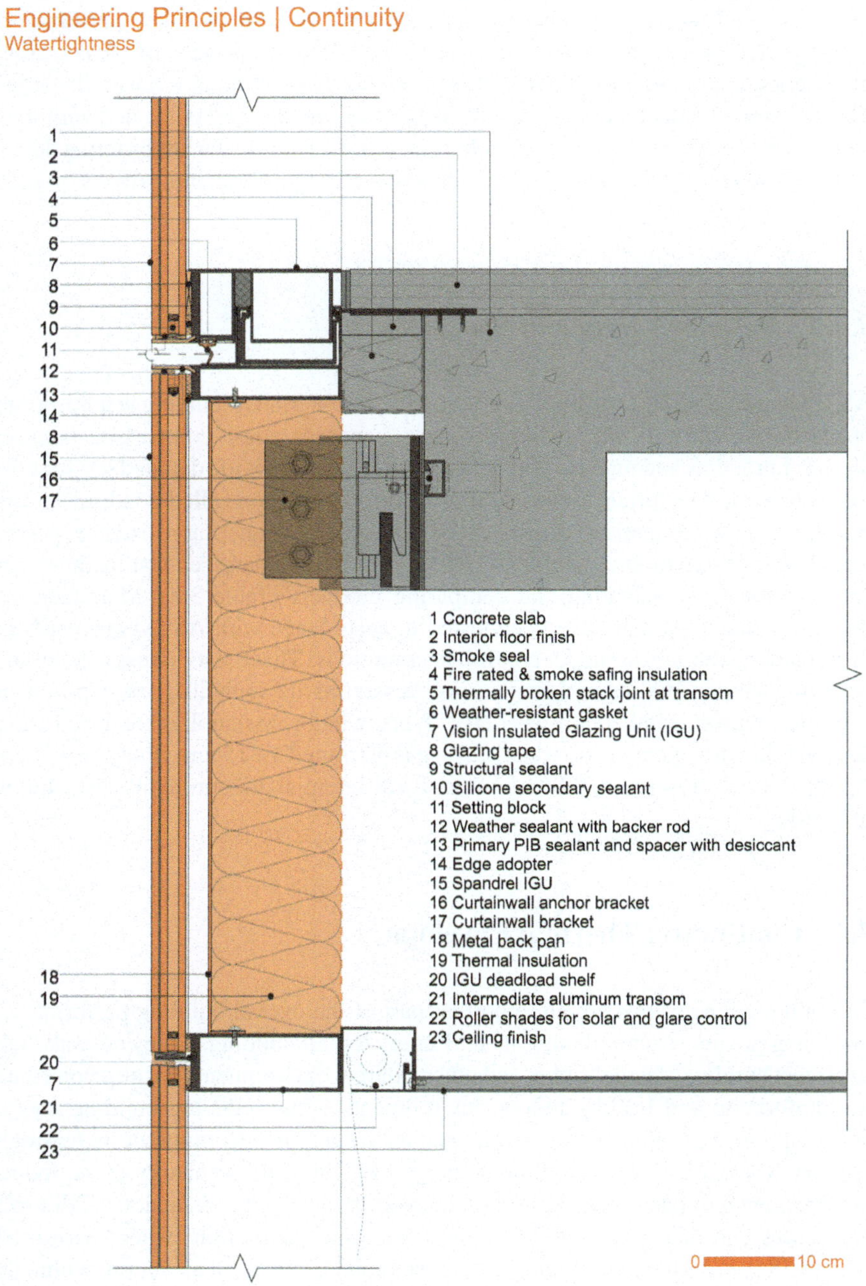

Fig. 3.4 Curtainwall components with water and air tightness performance

Table 3.8 ASTM testing for waterproofing products [247, 248]

ASTM test number	Test type
D-412	Tensile strength
D-413	Elongation
E-96	Moisture vapor transmission
D-822	Weathering resistance
C-501	Abrasion resistance
E-154	Puncture resistance
D-71	Specific gravity
D-93	Flash point
D-2240	Shore A hardness
E-42	Accelerated aging
E-119	Fire endurance
D-1149	Ozone resistance and weathering
C-67	Water repellency
E-514	Water permanence of masonry
C-109	Compressive strength
C-348	Flexural strength
D-903	Adhesion

Aqua Tower in Chicago exhibits its unique building form composed of undulating balconies that make the structure's exterior visually distinctive. However, this design introduced thermal bridging challenges through exposed balconies. The extended balcony overhangs became pathways for transferring heat through the connection between the exterior concrete balconies and the interior floor slabs, introducing condensation risks and increasing energy consumption [249]. Concrete balcony slabs without adequate thermal breaks result in increased heat loss, compromising the thermal performance of the building enclosure.

The approach to addressing thermal bridge at Aqua Tower demonstrates how aesthetics and energy performance can be balanced. The implemented solutions preserve the aesthetic character of the building while significantly enhancing its energy efficiency. Remediation efforts included the installation of thermal breaks at the connections between the exterior concrete balconies and the interior concrete floor slabs to reduce heat transfer. These thermal isolation strategies reduce energy flow between the balconies and the slabs reduce heat transfer. Additionally, operable windows and improved ventilation further reduced air-conditioning loads and enhanced indoor air quality.

Before its energy-efficient retrofitting, the Empire State Building experienced significant issues with thermal bridging, resulting in substantial energy loss. To address these inefficiencies, several energy-saving upgrades were implemented. According to Rocky Mountain Institute, a leading organization in energy efficient retrofits, the project included the installation of over 6000 insulated reflective barriers behind radiators to enhance interior thermal performance [250]. Additionally, more than 6500 windows were refurbished with advanced insulation technologies, significantly reducing heat loss through the building envelope and improving overall energy performance. These measures were critical in mitigating the effects of thermal bridging, increasing energy efficiency, and enhancing tenant comfort. The energy-efficiency upgrades have led to a reduction of up to 4000 metric tons of carbon emissions annually, comparable to the amount sequestered by a 750-acre pine forest. Additionally, the building's annual energy consumption is expected to decrease by 38%, translating to energy cost savings of approximately $4.4 million per year.

3.4.1 Design Criteria and Standards

Curtainwall façades are continuous across the front of the primary structure. Vision IGUs (insulated glass units) are attached to thermally broken metal or wood frames using mechanical fasteners and/or adhesives. The spandrel panel above the vision IGU includes insulation combined with a metal back pan, providing a higher thermal resistance (R-value). The curtainwall system is prefabricated, enabling improved quality control to minimize thermal bridging and ensure effective thermal insulation performance. Additionally, metal spacers within the glass panes of the IGU may be replaced with materials that have lower thermal conductivity to further enhance energy efficiency of the glazed system. Figure 3.5 illustrates key curtainwall components that enhance thermal insulation and energy efficiency.

U-Factor: The U-factor (Btu/h ft^2 °F or kWh/m^2 °K) of a glazing unit is related to space heating and cooling as well as occupant thermal comfort [251]. Energy transfer occurs due to temperature differences between indoor and outdoor environments, requiring HVAC systems to maintain target interior temperatures. The U-factor quantifies heat transfer through a building assembly by conduction, convection, and radiation mechanisms. Conduction and convection are the primary mechanisms of heat transfer, while radiative heat transfer plays a secondary role. Lower U-factor values indicate better thermal performance and greater energy efficiency. International Energy Conservation Code (IECC) defines the U-factor in compliance with NFRC 100 *Procedure for Determining Fenestration Product U-factors* and ASTM C1363-24 *Standard Test Method for Thermal Performance of Building Materials and Envelope Assemblies by Means of a Hot Box Apparatus.*

Solar Heat Gain Coefficient (280–2500 nm): The Solar Heat Gain Coefficient (SHGC) measures the fraction of solar heat that enters through a window relative to the total solar radiation incident on the window within the 280–2500 nm wavelength range [252]. The SHGC is calculated as the sum of solar energy that is directly transmitted, absorbed, and subsequently re-radiated into the interior space, reflecting the overall solar heat gain through the window system. In areas with excessive solar heat gain, windows should have a lower SHGC to minimize unwanted solar gain, whereas regions with limited sunlight benefit from windows with a higher SHGC to maximize beneficial solar gain. An SHGC of 1.0 indicates 100% transmission of

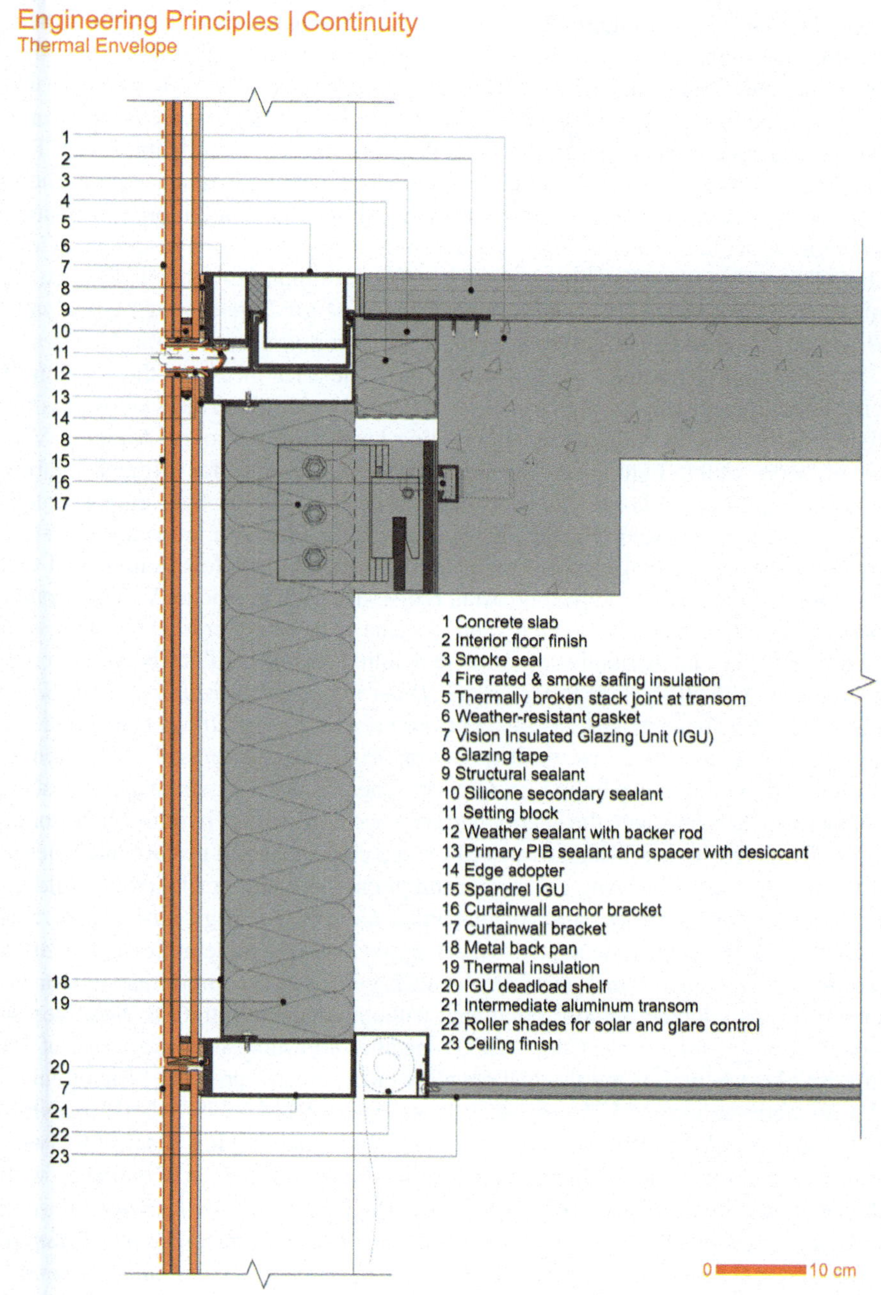

Fig. 3.5 Curtainwall components for improved thermal performance

incident solar energy, while an SHGC of 0.0 means no solar energy passes through the glazing. Low-emissivity (low-e) coated windows are energy efficient because they provide a low U-factor and reduced SHGC. When combined with shading devices that integrate building-integrated photovoltaics (BIPV) or vegetation screens, the overall energy and environmental performance of the system is significantly improved. This combination reduces cooling and heating demands, decreases reliance on artificial lighting, generates on-site renewable energy, and mitigates air pollution, contributing to a more sustainable building envelope. The International Energy Conservation Code (IECC) specifies SHGC requirements depending on climate zones. SHGC value are measured in accordance with NFRC *200 Procedure for Determining Fenestration Product Solar Heat Gain Coefficients and Visible Transmittance at Normal Incidence.*

Visible Light Transmittance (380–780 nm): The VLT measures the fraction of visible sunlight passing through a window at normal incidence to the window. Visible light transmission directly affects daylighting levels inside a room. A higher VLT means more visible light enters the space, reducing the need for artificial lighting and lowering energy costs. A lower VLT means less visible light enters, potentially requiring increased reliance on electric lighting. However, transmitted visible light also carries thermal energy, which can affect thermal comfort. A high VLT can increase cooling loads, thereby increasing HVAC demand. Balancing VLT with other performance factors such as SHGC and U-value is therefore critical for achieving energy-efficient and thermally comfortable building design. VLT can be measured in accordance with NFRC 200 *Procedure for Determining Fenestration Product Solar Heat Gain Coefficients and Visible Transmittance at Normal Incidence.* Table 3.9 explains the maximum U-factor and maximum SHGC requirements of vertical fenestration and skylights depending on climate zones. Table 3.11 summarizes testing standards used to evaluate the energy performance attributes of building enclosures.

Air Leakage: The air leakage rate (CFM) is a critical energy-performance metric that impacts overall building energy consumption. Gaps and cracks in the building envelope are the primary cause of unwanted air leakage. High air-leakage rates increase HVAC energy demand to meet target indoor temperature and relative humidity levels for thermal comfort. Air leakage can also compromise indoor air quality by introducing outdoor pollutants without adequate filtration. The International Energy Conservation Code (IECC) specifies allowable air-leakage rate of 0.4 cfm/ft^2 (2.0 L/s m^2) of the thermal envelope area under a pressure differential of 0.3 in. water column (75 Pa), in accordance with ASTM E779, ANSI/RESNET/ICC 380, ASTM E3158, or ASTM E1827. Different building enclosure types can be tested under field conditions and/or laboratory settings in accordance with ASTM E283 *Standard Test Method for Determining Rate of Air Leakage Through Exterior Windows, Skylights, Curtain Walls, and Doors Under Specified Pressure Differences Across the Specimen*, NFRC 400 *Procedure for Determining Fenestration Product Air Leakage*, and AMMA/WDMA/CSA101/I.S.2/A440-2022 *North American Fenestration Standard/Specification For Windows, Doors, and Skylights* (Table 3.10).

Table 3.9 IECC's Table C402.4 maximum U-factor and SHGC requirements of building envelope fenestration [253]

Climate zone	0 and 1		2		3		4 (except marine)		5 and marine 4		6		7	
Vertical fenestration														
U-factor														
Fixed fenestrations	0.5		0.45		0.42		0.36		0.36		0.34		0.29	
Operable fenestrations	0.62		0.6		0.54		0.45		0.45		0.42		0.36	
Entrance doors	0.83		0.77		0.68		0.63		0.63		0.63		0.63	
SHGC														
Projection factor (PF)	F	O	F	O	F	O	F	O	F	O	F	O	F	O
PF < 0.2	0.23	0.21	0.25	0.23	0.25	0.23	0.36	0.33	0.38	0.33	0.38	0.34	0.4	0.36
0.2 ≤ PF ≤ 0.5	0.28	0.25	0.3	0.28	0.3	0.28	0.43	0.4	0.46	0.4	0.46	0.41	0.48	0.43
PF > 0.5	0.37	0.34	0.40–	0.37	0.4	0.37	0.58	0.53	0.61	0.53	0.61	0.54	0.64	0.58
Skylights														
U-factor	0.7		0.65		0.55		0.5		0.5		0.5		0.44	
SHGC	0.3		0.3		0.3		0.4		0.4		0.4		NR	

F fixed fenestration, *O* operable fenestration

Table 3.10 IECC's Table C402.5 maximum air leakage rate for fenestration assemblies [254]

Fenestration assembly	Maximum rate (cfm/ft^2)	Test procedure
Windows	0.20[a]	AAMA/WDMA/CSA101/I.S.2/ A440 or NFRC 400
Sliding doors	0.20[a]	
Swinging doors	0.20[a]	
Skylights—with condensation weepage openings	0.30	
Skylights—all other	0.20	
Curtain walls	0.06	NFRC 400 or ASTM E283 at 1.57 psf (75 Pa)
Storefront glazing	0.06	
Commercial glazed swinging entrance doors	1.00	
Power-operated sliding doors and power-operated folding doors	1.00	
Revolving doors	1.00	
Garage doors	0.40	ANSI/DASMA 105, NFRC 400, or ASTM E283 at 1.57 psf (75 Pa)
Rolling doors	1.00	
High-speed doors	1.30	

For SI: 1 cfm = 0.47 L/s; 1 ft.^{2} = 0.093 m^2
[a] The maximum air leakage rate for windows, sliding doors, swinging doors, and skylights is permitted to be increased to 0.3 cfm per square of fenestration or door area when tested in accordance with AAMA/WDMA/CSA 101/I.S.2/A440 at 6.24 psf (300 Pa)

Vapor Permeability: Vapor permeability is a critical factor that requires special attention in the design and selection of vapor retarders for building enclosures. Moisture from condensation or other vapor sources can significantly impact the thermal resistance of insulation, and the effect of moisture must be evaluated for each specific material layer with building envelope assemblies. Some materials with large air spaces may experience minimal performance degradation if the moisture content is less than approximately 10% by weight, while other materials exhibit approximately linear reduction in thermal performance as moisture content increases [255] (Table 3.11).

3.4.2 *Heat Air Moisture (HAM) Performance Metrics*

Thermal Insulation: The overall values of thermal resistance, or R-values, of a wall assembly are calculated with the assumptions that building materials and their components exist under ideal conditions. This implies that each insulating material has a consistent nominal thickness, and its corresponding thermal resistance is uniform throughout. Other assumptions include uniform thicknesses of air spaces, even surface temperatures, and the absence of moisture effects. In addition, it is

Table 3.11 Testing standards related to energy attributes of building enclosures [256–258]

Testing standards related to energy attributes of building enclosures
R-value/U-factor related standards
ASTM C1363 Standard Test Method for Thermal Performance of Building Materials and Envelope Assemblies by Means of a Hot Box Apparatus
ASTM C168 Terminology Relating to Thermal Insulation
ASTM C177 Test Method for Steady-State Heat Flux Measurements and Thermal Transmission Properties by Means of the Guarded-Hot-Plate Apparatus
ASTM C518 Test Method for Steady-State Thermal Transmission Properties by Means of the Heat Flow Meter Apparatus
ASTM C739 Specification for Cellulosic Fiber Loose-Fill Thermal Insulation
ASTM C764 Specification for Mineral Fiber Loose-Fill Thermal Insulation
ASTM C870 Practice for Conditioning of Thermal Insulating Materials
ASTM C1045 Practice for Calculating Thermal Transmission Properties Under Steady-State Conditions
ASTM C1058 Practice for Selecting Temperatures for Evaluating and Reporting Thermal Properties of Thermal Insulation
ASTM C1130 Practice for Calibration of Thin Heat Flux Transducers
ASTM C1199 Test Method for Measuring the Steady-State Thermal Transmittance of Fenestration Systems Using Hot Box Methods
ASTM C1224 Specification for Reflective Insulation for Building Applications
ASTM C1371 Test Method for Determination of Emittance of Materials Near Room Temperature Using Portable Emissometers
ASTM C1373 Practice for Determination of Thermal Resistance of Attic Insulation Systems Under Simulated Winter Conditions
ASTM E177 Practice for Use of the Terms Precision and Bias in ASTM Test Methods
ASTM E230 Specification for Temperature-Electromotive Force (EMF) Tables for Standardized Thermocouples
ASTM E691 Practice for Conducting an Interlaboratory Study to Determine the Precision of a Test Method
ASTM E903 Test Method for Solar Absorptance, Reflectance, and Transmittance of Materials Using Integrating Spheres
ASTM E1423 Practice for Determining Steady State Thermal Transmittance of Fenestration Systems
ASTM E1424 Test Method for Determining the Rate of Air Leakage Through Exterior Windows, Skylights, Curtain Walls, and Doors Under Specified Pressure and Temperature Differences Across the Specimen
ASHRAE Handbook of Fundamentals, Latest Edition, American Society of Heating, Refrigerating and Air Conditioning Engineers, Inc.
ISO Standard 8990 Thermal Insulation Determination of Steady State Thermal Properties—Calibrated and Guarded Hot Box, ISO 8990-1994(E)
ISO Standard 12567 Thermal Performance of Windows and Doors—Determination of Thermal Transmittance by Hot Box Method, ISO 12567-2000

(continued)

Table 3.11 (continued)

Testing standards related to energy attributes of building enclosures
SHGC and VLT related standards ASTM E903-20 Standard Test Method for Solar Absorptance, Reflectance, and Transmittance of Materials Using Integrating Spheres ASTM E275 Practice for Describing and Measuring Performance of Ultraviolet and Visible Spectrophotometers ASTM E424 Test Methods for Solar Energy Transmittance and Reflectance (Terrestrial) of Sheet Materials ASTM E490 Standard Solar Constant and Zero Air Mass Solar Spectral Irradiance Tables ASTM E772 Terminology of Solar Energy Conversion ASTM E971 Practice for Calculation of Photometric Transmittance and Reflectance of Materials to Solar Radiation ASTM E1084 Test Method for Solar Transmittance (Terrestrial) of Sheet Materials Using Sunlight ASTM E1175 Test Method for Determining Solar or Photopic Reflectance, Transmittance, and Absorptance of Materials Using a Large Diameter Integrating Sphere ASTM E2554 Practice for Estimating and Monitoring the Uncertainty of Test Results of a Test Method Using Control Chart Techniques ASTM G173 Tables for Reference Solar Spectral Irradiances: Direct Normal and Hemispherical on 37° Tilted Surface ASTM G197 Table for Reference Solar Spectral Distributions: Direct and Diffuse on 20° Tilted and Vertical Surfaces Federal Test Method Standard No. 141, Method 6101 ASHRAE Standard 74-1988 CIE 38 Radiometric and Photometric Characteristics of Materials and their Measurement CIE 44 Absolute Methods for Reflection NIST SP 250-69 Regular Spectral Transmittance
Air leakage standards ASTM D8052/D8052M Test Method for Quantification of Air Leakage in Low-Sloped Membrane Roof Assemblies ASTM E283/E283M Test Method for Determining Rate of Air Leakage Through Exterior Windows, Skylights, Curtain Walls, and Doors Under Specified Pressure Differences Across the Specimen ASTM E631 Terminology of Building Constructions ASTM E779 Standard Test Method for Determining Air Leakage Rate by Fan Pressurization ASTM E1677 Specification for Air Barrier (AB) Material or Assemblies for Low-Rise Framed Building Walls ASTM E1827 Standard Test Methods for Determining Airtightness of Buildings Using an Orifice Blower Door ASTM E2178 Standard Test Method for Determining Air Leakage Rate and Calculation of Air Permeance of Building Materials ASTM E2357 Test Method for Determining Air Leakage Rate of Air Barrier Assemblies ANSI/RESNET/ICC 380 Standard for Testing Airtightness of Building Enclosures, Airtightness of Heating and Cooling Air Distribution Systems, and Airflow of Mechanical Ventilation Systems AMMA/WDMA/CSA 101/I.S.2/A440-2022 North American Fenestration Standard/ Specification for Windows, Doors, and Skylights

assumed that wall installation is carried out strictly in accordance with the specified construction details. However, as noted in Building Materials and Structures Report (BMS) 151 by the National Institute of Standards and Technology (NIST), there are insulated wall assemblies, for which measured R-values do not correspond to calculated values. This highlights the need to account for real-world performance and the efficiency of building envelopes under actual field conditions.

Thermal Inflection: The thermal inflection point is defined as the insulation thickness at which energy savings begin to diminish or where additional insulation yields diminishing or negative values. Exceeding the optimal thickness of insulation produces the opposite effect, reducing overall energy efficiency. A similar situation implies that insulation thickness influences the peak heating and peak cooling loads of the HVAC systems. Peak heating loads represent the energy demand during extreme cold days, whereas peak cooling loads represent the maximum energy demand during hot weather days. The use of adequate insulation ensures that these peak energy loads are minimized, thereby enabling energy-efficient and cost-effective HVAC operation. Climate zones affect optimal insulation levels, which should be selected to balance energy efficiency and occupant comfort.

Al-Sanea et al. [259] studied optimal R-values for wall construction in Saudi Arabian cities with varying climates. Using red clay bricks and concrete for one- and two-layered wall assemblies, their computer model found ideal R-values of 2.00–2.90 m^2 °K/W for Riyadh and Jeddah (hot and dry climates) and 1.34–1.99 m^2 °K/W for Abha (moderate climate). The calculated payback period at these optimal R-values, considering inflation and discount rates, was found to range from approximately three to ten years.

Bojic et al. [260] studied how building envelope design affects space cooling loads in hot climates, focusing on residential structures with extensive evening and nighttime air-conditioning use. Their findings suggest that increasing thermal insulation and thermal mass in external walls can have counteracting effects. Increasing thermal mass may increase heat storage capacity, leading to higher cooling needs during evening hours. However, increasing insulation can reduce cooling loads by minimizing heat transfer through external walls with thickness of approximately 250 mm.

Bojic and Yik [261] explored the relationship between the design of external walls, internal partitions, and thermal insulation layers and the space cooling loads of residential flats in Hong Kong. Their findings showed that insulating both the building envelope and interior partitions can reduce annual space cooling loads by up to 38%. However, the number and placement of insulation layers may increase the peak cooling demand by up to 19% or decrease it by up to 16%, depending on assembly configuration. In unfavorable scenarios, significant increases in peak cooling demand—exceeding 60%—were observed when thermal capacity in the envelope and partitions was reduced.

Thermal Bridge: Thermal bridges significantly degrade a building's thermal performance. They create pathways for heat flow, undermining insulation effectiveness. This unintended heat transfer can reduce effective R-values, lower energy efficiency, and promote mold formation. Addressing thermal bridges through continuous insulation, thermal breaks, and other mitigation strategies is therefore crucial to maintaining intended thermal performance and acceptable indoor air quality.

Energy performance was conducted to understand the impact of thermal bridging on energy consumption both in new building and retrofit scenarios using hourly energy simulations [262]. Findings indicated that the energy consumption of a villa constructed with insulated blocks in a typical Dubai configuration, but with uninsulated reinforced concrete frame, was approximately 24.5% higher than that of an identical villa with full perimeter external 50-mm expanded polystyrene (EPS) insulation over a prefabricated concrete block. Moreover, insulating only the reinforced concrete frame resulted in approximately 23% energy savings compared to the base scenario, emphasizing the significance of addressing thermal bridging to improve energy efficiency. Another study investigated the dynamic impact of thermal bridges on residential building energy efficiency using three analytical approaches: the U-value method, the equivalent wall method, and the direct 2D/3D numerical modeling using WUFI plus, a whole-building heat, air, and moisture (HAM) simulation tool [263]. The findings highlighted the importance of accounting for thermal bridges in energy-efficiency assessments in both cooling- and heating-dominant climates. In cold climates, including thermal bridges resulted in 8–13% higher annual heating loads when modeled dynamically, compared with the equivalent U-value and equivalent-wall methods. In hot climates, thermal bridges estimated by dynamic modeling led to a 20% increase in annual cooling demand, while the U-value and equivalent-wall methods underestimated cooling demand by 17% and 14% respectively, compared to dynamic modeling results.

Solar Stress: During the early design phase, energy modeling techniques can be applied to determine the most appropriate façade design to conserve energy in regions where cooling is a dominant demand. In terms of the cooling load component, studies have established that the solar heat gain coefficient (SHGC) is the most important glazing characteristic influencing building energy consumption. The findings suggested that improved cooling-load reduction could be achieved with an SHGC of below 0.3, and that for lowering overall energy consumption, a lower SHGC is more energy savings than a low U-value [264]. Further analysis indicated that energy use intensity (EUI) correlates more strongly with SHGC than with U-value, with the R^2 (coefficient of determination) values of 0.98 for SHGC and 0.66 for U-value, respectively. Lee et al. assessed various performance parameters of windows in determining energy efficiency of office buildings in a hot-humid region [265]. Their energy modeling revealed that SHGC is the most important variable among other factors affecting the heating and cooling loads, followed by U-value and Visible Light Transmittance (VLT). For buildings with high window-wall-ratios (WWR), SHGC was shown to be particularly critical. In buildings with above-average WWRs, appropriate combinations of low U-value and optimized SHGC are important to improve the overall energy efficiency of window systems.

Daylight Illumination: Dynamic windows could reduce energy consumption across various climates by leveraging their adaptable properties of VLT and SHGC. In hot and dry climates such as Phoenix, Arizona, electrochromic windows have proven effective, where building energy use was reduced by approximately 46% compared with single-pane clear glazing [266]. Another study investigated the functioning of suspended particle device (SPD) glazing, focusing specifically on hot desert climates [267]. The findings indicated that the use of switchable VLT and SGHC helps reduce building energy consumption while improving visual comfort. More specifically, dynamic glazing resulted in a 29.1% reduction in cooling energy consumption, highlighting the energy-efficiency benefits of this technology in hot desert regions.

Air Tightness: Preventing unwanted air infiltration through building envelopes is essential for reducing energy consumption and maintaining humidity control. Building envelope air tightness has been shown to significantly reduce energy consumption in hot, humid climates [268]. Three scenarios were evaluated: (1) a baseline condition in which the entire house is not airtight (Case 1), (2) a condition in which only the air-conditioned room was airtight (Case 2), and (3) a condition in which the entire house was airtight (Case 3). Enhancing airtightness only in air-conditioned rooms (Case 2) led to a 25% reduction in energy usage and a 50% reduction in latent load. When every room was air-conditioned and the entire house was airtight (Case 3), the energy usage reduction was limited to approximately 5–15% due to the increased cooling area. Reducing air exchange rates between air-conditioned rooms and the outdoor environment was identified as an efficient energy-saving strategy.

Another study investigated the impact of home airtightness on indoor temperature conditions in hot summer and cold winter climates. Airtightness significantly affected heating energy consumption but had minimal impact on cooling energy consumption [269]. The study recommended higher airtightness levels for cities at higher latitudes and in cold climate zones with significant winter temperature differences. Cities with mild winter temperature differences and low wind pressure may not require airtightness levels as high as those recommended for cold climate regions.

Chapter 4
Hot Climate Tall Tower Case Studies

Abstract This chapter discusses three case studies and façade assembly details of tall buildings located in hot and humid climates: Shenzhen Energy Mansion in Shenzhen, China, MahaNakhon Tower in Bangkok, Thailand, and Le Nouvel KLCC in Kuala Lumpur, Malaysia. It aims to analyze and evaluate how these tall building enclosures implement sustainable design and engineering principles and address climatic, environmental, and vertical urbanization challenges. The Shenzhen Energy Mansion demonstrates how folded curtainwall systems improve building energy efficiency and indoor environmental quality. The folded curtainwall shows the interplay between transparency and opacity to control daylighting, solar heat gain, glare, and natural ventilation. The MahaNakhon Tower in Bangkok features pixelated and cassette façade systems. The pixelated façade creates green garden balconies in a spiraling configuration, supporting urban connectivity and biophilic design. The cassette façade adds a clean, sleek exterior expression, articulated by precise shadow-line reveals that enhance facade rhythm and visual refinement across the building envelope. These façades contribute to regulating the building's microclimate through shading and enhanced natural ventilation, achieving a balanced integration of form, function, and architectural expression. The Le Nouvel KLCC in Kula Lumpur exemplifies biophilic design through the integration of nature within the curtainwall façade, promoting improved air quality, thermal comfort, solar shading, and urban biodiversity. Collectively, these continuity-framework case studies highlight the importance of sustainable design and engineering principles for tall buildings in hot climates, showing how these approaches can address rapid urbanization challenges while enhancing environmental sustainability, urban ecological systems, and occupant well-being.

Keywords Shenzhen Energy Mansion · MahaNakhon Tower · Le Nouver KLCC · Sustainable tall buildings · High-performance facades · Bioclimatic design strategies · Biophilic design · Hot humid climate design

K. H. Kim, *Continuity in Sustainable Facades*,
https://doi.org/10.1007/978-3-032-19841-9_4

4.1 Shenzhen Energy Mansion, Shenzhen, China, 2018

> Shenzhen Energy Mansion appears as a subtle mutation of the classic skyscraper and exploits the building's interface with the external elements: sun, daylight, humidity and wind to create maximum comfort and quality inside. A natural evolution that looks different because it performs differently. Bjarke Ingels, Founding Partner, BIG

Shenzhen Energy Mansion has two towers linked by a podium at its base, standing 220 m (721.8 ft) tall to the north and 120 m (393.7 ft) tall to the south in Shenzhen, China. The program consists of office towers, and a podium housing lobbies, cafeteria spaces, conference center, and exhibition areas, covering an area of 96,000 m^2 (1,033,334 ft^2). Developed by Shenzhen Energy Company, the complex features a sinuous folded façade comprised of transparent and opaque elements. The folded curtainwalls are articulated in an undulating form, providing multiple functions while helping the high rise achieve iconic sustainable architecture. For example, the folded curtainwall's opaque element provides effective thermal insulation reducing heat transfer, solar heat gain, and glare, whereas the folded curtainwall's transparent element encourages daylight penetration into interior spaces, natural ventilation through operable windows, and clear views from office spaces. The transparent layer of the folded curtainwall is oriented primarily toward the north, creating interior spaces abundant in diffuse daylight. This effect is achieved by admitting northern light and reflecting direct sunlight off the folded curtainwall and between the interior spaces. Even when the sun is positioned directly to the east or west, the folded curtainwall ensures that most solar rays are reflected away from the vision window. These high-performance building enclosures reduce air conditioning energy consumption by 20% [270]. Photovoltaic panels on the roof generate electricity, while rainwater harvesting systems allow water reuse for cooling towers and landscape irrigation, among other building uses (Figs. 4.1 and 4.2).

Shenzhen's rapid urbanization has resulted from its economic growth over the last few decades. While economic transformation drives urban growth, it also creates numerous environmental challenges including air pollution and increased water and electricity demand [271]. Better views can be enjoyed from higher floors due to taller structures, which also can also help indoor air quality by allowing air circulation. Since many towers today are built with sustainable technologies such as high-performance glass windows and energy-saving lighting and HVAC systems, they tend to consume less electricity while providing comfortable indoor conditions. Tall buildings create iconic skylines that are visually pleasing and promote tourism in cities, attracting visitors from around the world who come to experience distinctive architecture and urban environments.

Advantages of constructing taller buildings in places like Shenzhen include increased usable floor area, proximity to work and urban infrastructure, sustainability benefits from densified development, and iconic skylines that contribute to increased tourism, economic prosperity, global recognition, among other benefits. Constructing additional skyscrapers can be advantageous both economically and ecologically when properly planned, provided that appropriate architectural and

Fig. 4.1 Exterior view of Shenzhen Energy Mansion, Shenzhen, China (drawings reproduced courtesy of Bjarke Ingels Group)

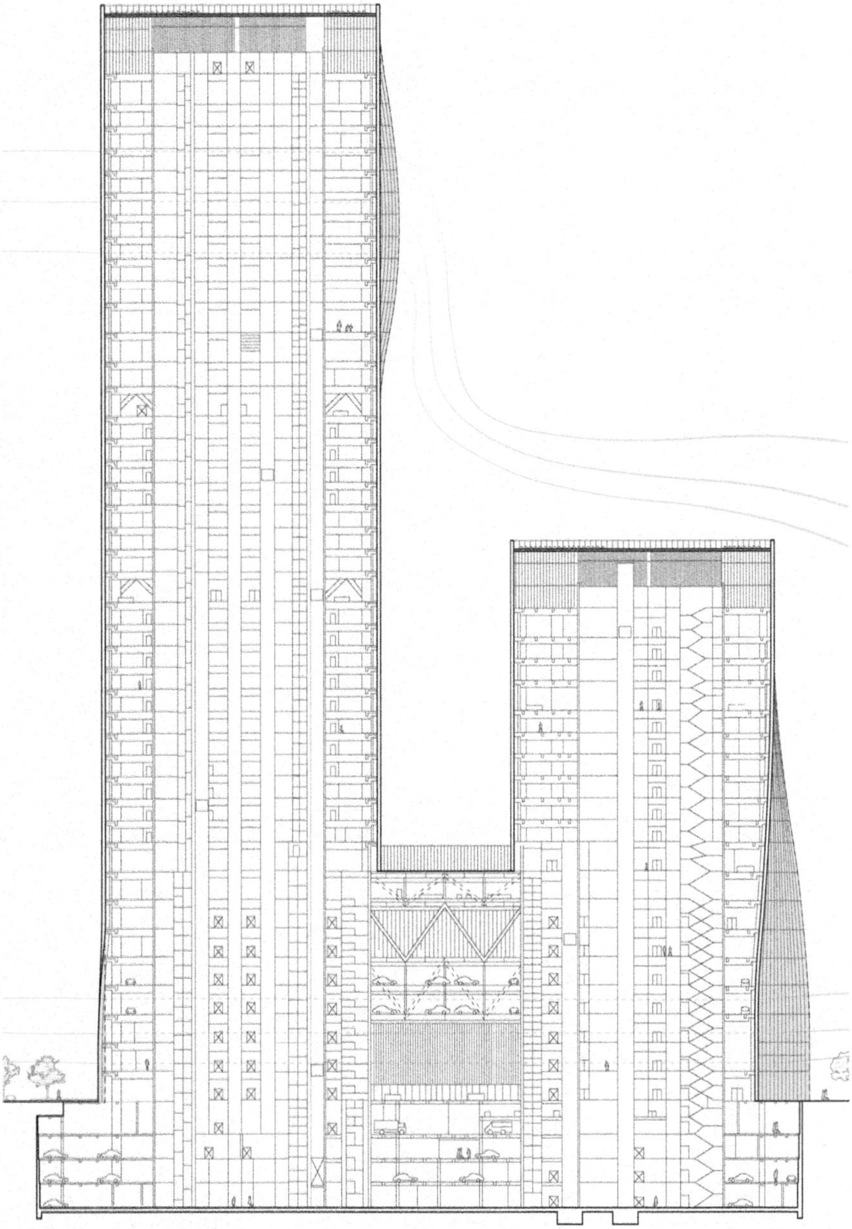

Fig. 4.2 Building section view of Shenzhen Energy Mansion, Shenzhen, China (drawings reproduced courtesy of Bjarke Ingels Group)

zoning regulations are implemented to address requirements for sustainable built environments, including energy efficiency, water conservation, the use of sustainable materials, wildlife habitats, water quality, and air pollution. One design strategy that can help reduce the environmental impact of tall buildings is the integration of green spaces such as sky gardens, terraced gardens, or rooftop terraces. These are not only architecturally appealing but also help mitigate dense urban environments while improving air quality and promoting biodiversity through increased vegetation [272]. In addition, they enable occupants to enjoy outdoor activities such as gardening and access to fresh air above street level, away from traffic-related pollution.

Shenzhen, located in southeastern China's Guangdong province, was a relatively small town until 1979, when it became one of China's Special Economic Zones. This designation transformed the city into a major urban center, with an area of 1991 km^2 (768.7 mi^2) and a population exceeding 13 million as of 2024 [273]. Shenzhen's proximity to Hong Kong makes it ideal for international trade, benefiting from Hong Kong's strategic location, political stability, low corruption rates, and well-developed legal frameworks—advantages that were not historically common in many other Chinese cities [274]. Its landscape combines contemporary architecture with green infrastructure, positioning Shenzhen as a global innovation center, as evidenced by its ecologically oriented urban growth [275]. Shenzhen has what is referred to as a warm, humid subtropical climate (Köppen: Cfa), which means hot summers coupled with wet weather patterns throughout the year while winters are slightly cool [276] (Figs. 4.3, 4.4, 4.5, 4.6, 4.7, 4.8, 4.9, 4.10, 4.11, 4.12, 4.13, 4.14, 4.15, 4.16, 4.17, 4.18, 4.19 and 4.20).

4.2 MahaNakhon, Bangkok, Thailand, 2016

> We see architecture as process and the building as a dialogue - between large and small scale, symbolism and activity, the city and its inhabitants, inside and outside. Bangkok is a city of many contradictions - and of an untamable energy. Buro Ole Sheeren

The MahaNakhon complex, designed by Buro Ole Sheeren, intimately engages with Bangkok from its base to its pinnacle. The MahaNakhon development comprises 77 stories, reaching a height of 313 m (1026.9 ft) and covering a total floor area of 150,000 m^2 (1,614,585 ft^2). This mixed-use project by PACE Development Co., Ltd. integrates a diverse array of functions. Residential space accounts for 67,000 m^2 (721,200 ft^2), supported by 4400 m^2 (47,400 ft^2) of residential amenities. The hotel component includes 150 rooms with a total floor area of 16,600 m^2 (178,600 ft^2), complemented by 11,600 m^2 (124,900 ft^2) of hotel amenities. Retail functions occupy 5900 m^2 (63,500 ft^2), while the sky bar encompasses 3500 m^2 (37,700 ft^2). Mechanical, electrical, and plumbing (MEP) facilities account for 5200 m^2 (56,000 ft^2), and on-site parking areas total 10,400 m^2 (111,900 ft^2). In addition, a retail cube provides 9000 m^2 (96,900 ft^2) of retail space, along with 1000 m^2 (10,800 ft^2) dedicated to MEP facilities and 5400 m^2 (58,100 ft^2) allocated for parking. Off-site parking further

Fig. 4.3 Shenzhen Energy Mansion, Shenzhen, China (2018), and sustainable façade applications

Location	Shenzhen Energy Mansion, China
Köppen Climate Classification	Humid Subtropical Climate (Köppen: Cfa)
Completion Date	2018
Building Owner/Primary Tenant	Shenzhen Energy
Architects & Design Teams	Bjarke Ingels Group
Number of Floors	42 (north) and 19 (south)
Gross Floor Areas	96,000 m²

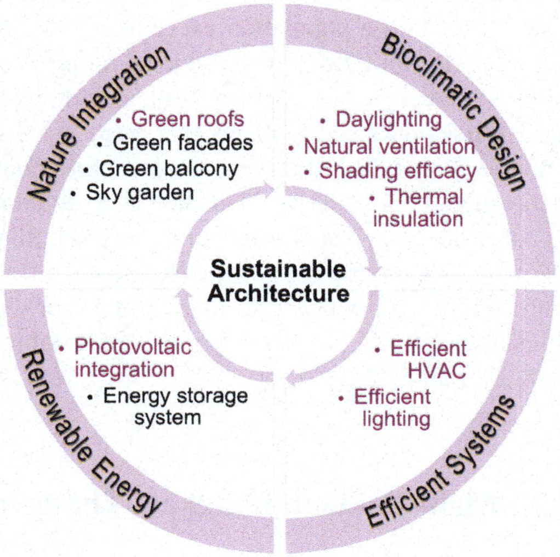

contributes 8300 m^2 (89,300 ft^2) to the overall development. Its distinctive design has made MahaNakhon an iconic structure within Bangkok's cityscape and a symbol of modernity throughout Thailand. According to the architect, MahaNakhon—translated as the "Great Metropolis," is conceived as a prototype that fosters urban and social interaction by actively engaging with the city. Its pixelated and eroded appearance fabric creates a dynamic, distinctive architectural presence, connecting with the surrounding urban fabric through a three-dimensional wrapping of pixels that spiral around the tower's height, metaphorically revealing the building's internal life and scale to the city [277] (Figs. 4.21 and 4.22).

Pixelated façades offer a dynamic urban experience by carving outdoor spaces into the building mass. The pixelated façade often features protruding and recessed elements that form terraces, allowing them to be used for greenery, social spaces, or private balconies, thereby contributing to the building's sustainability and aesthetic appeal. This approach provides accessible communal and recreational areas within high-rise buildings, enhancing user experience and spatial diversity. MahaNakhon's

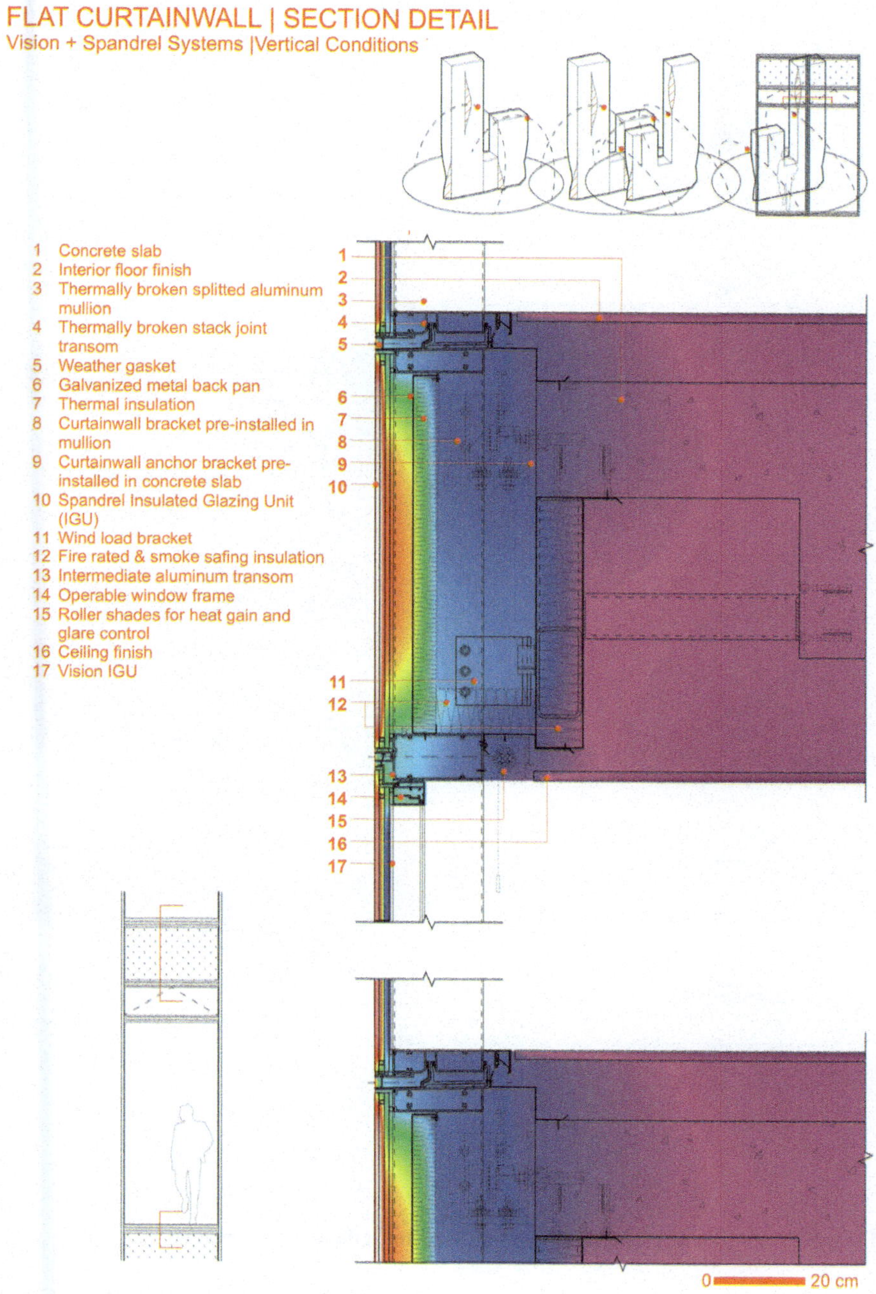

Fig. 4.4 Section detail of the flat curtainwall at the vertical condition

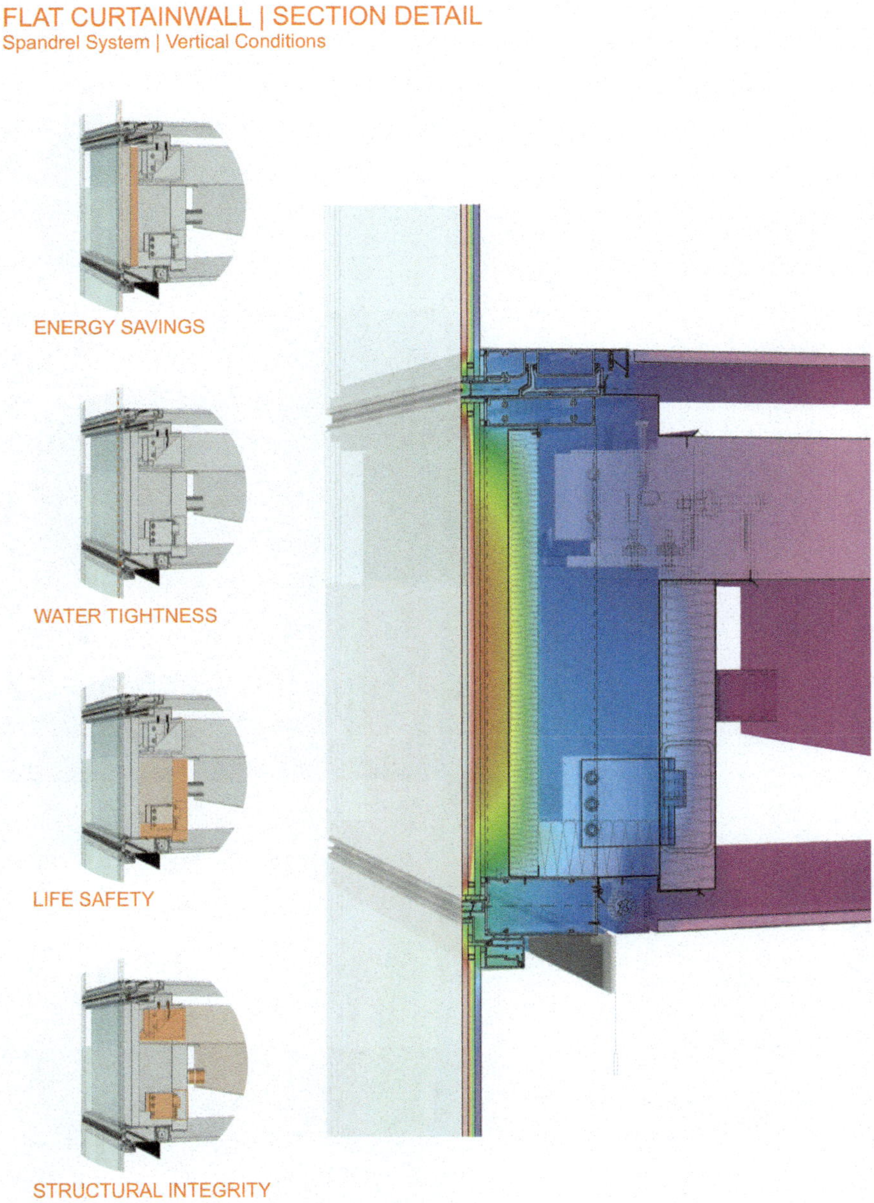

Fig. 4.5 Section detail of the flat curtainwall spandrel at the vertical condition

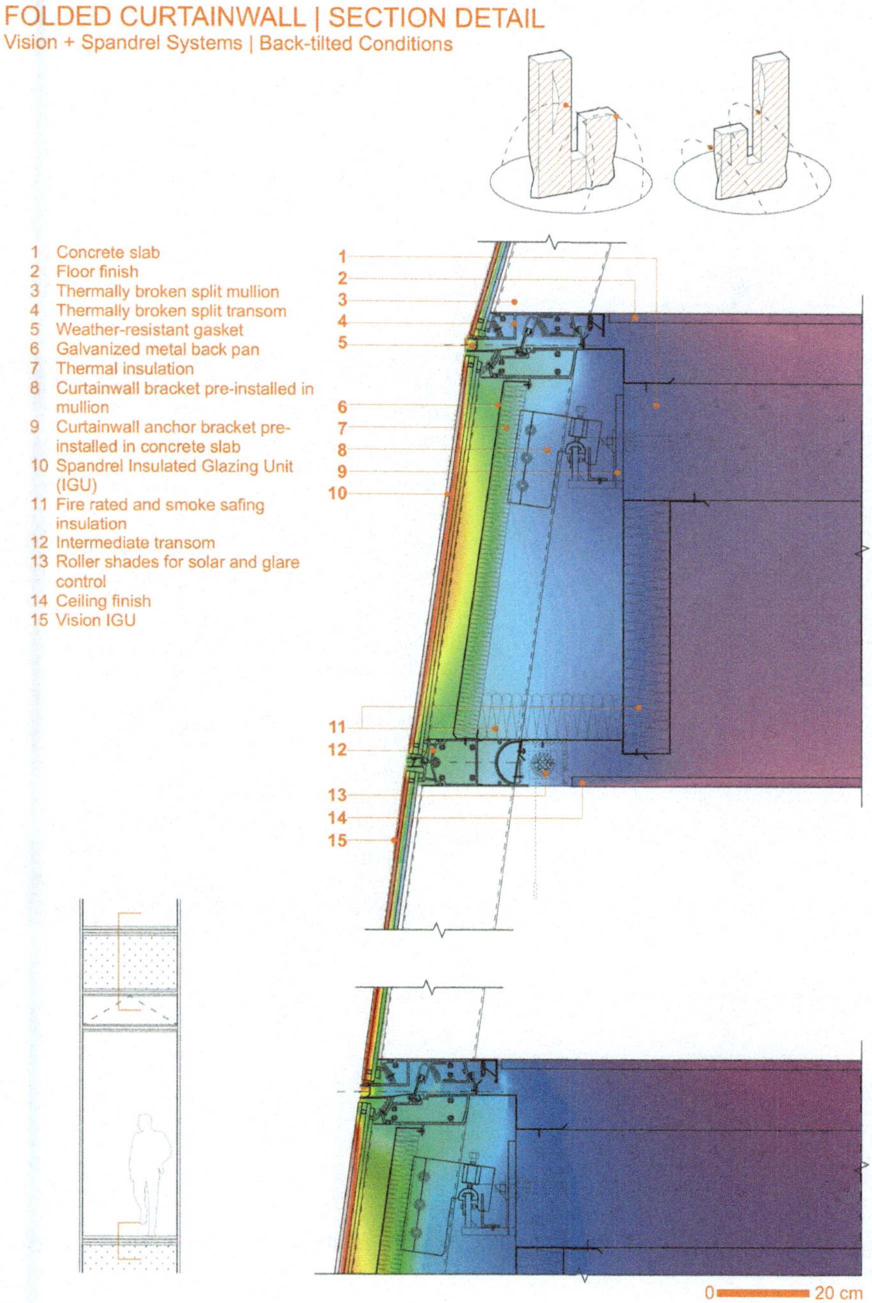

Fig. 4.6 Section detail of the folded curtainwall at the back-tilted condition

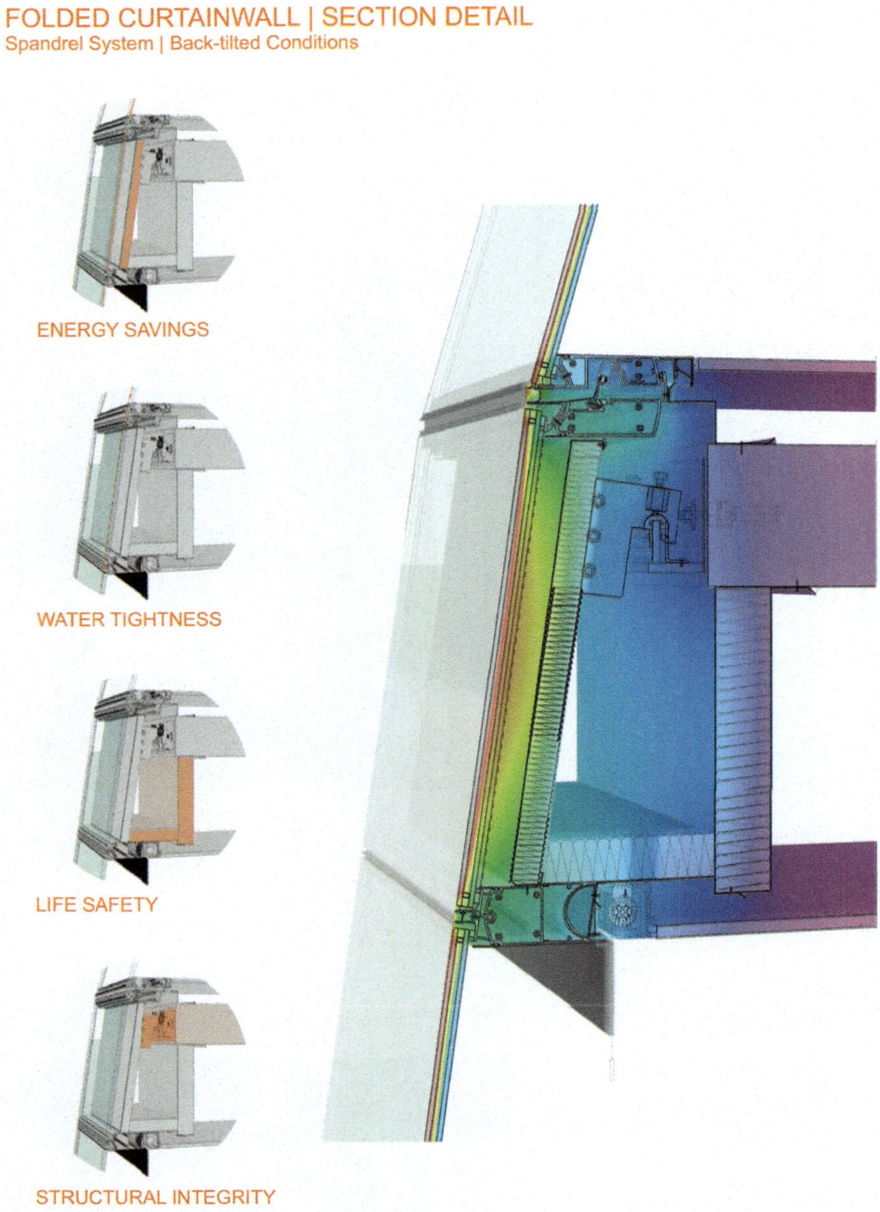

Fig. 4.7 Section detail of the folded curtainwall spandrel at the back-tilted condition spandrel panel at the parapet level

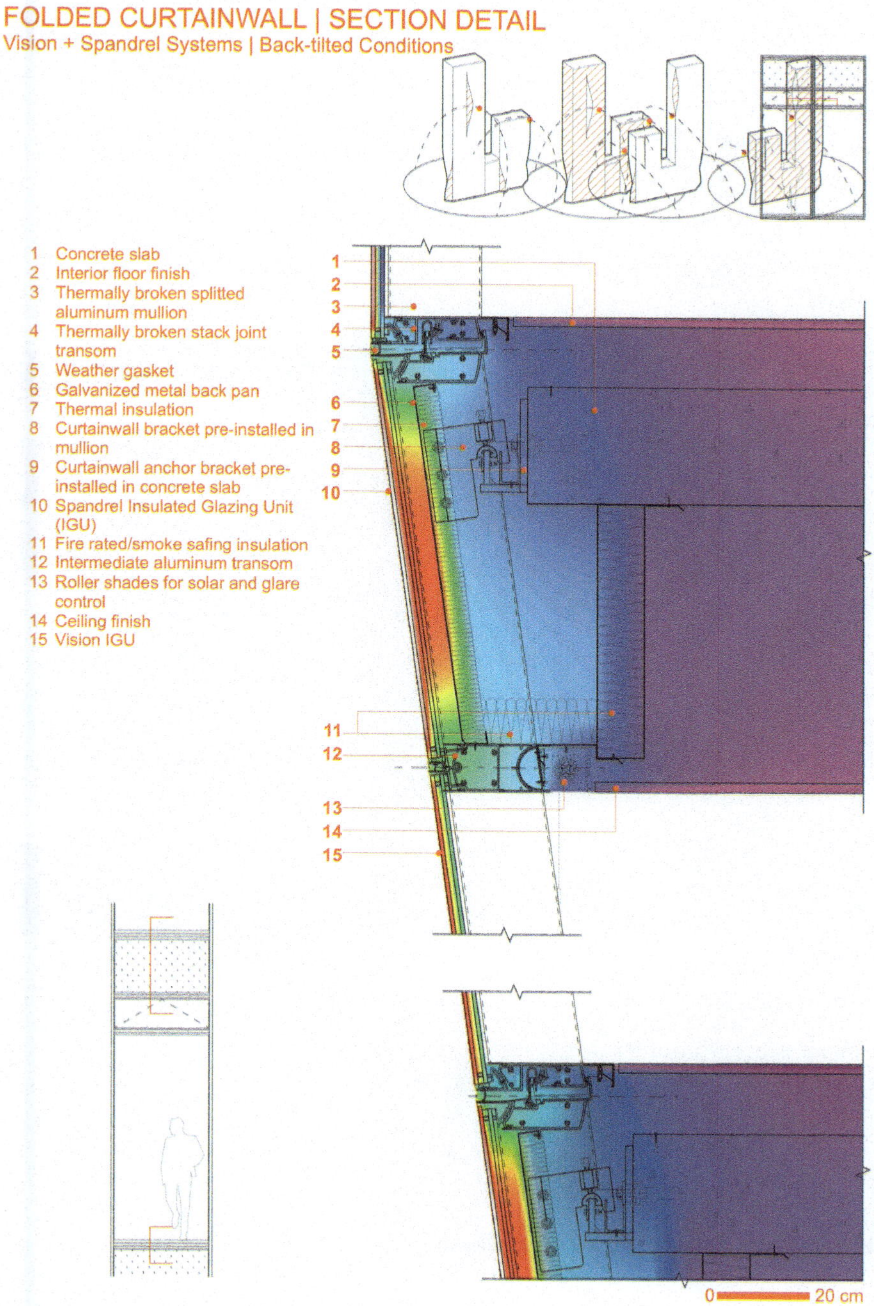

Fig. 4.8 Section detail of the folded curtainwall at the underslung condition

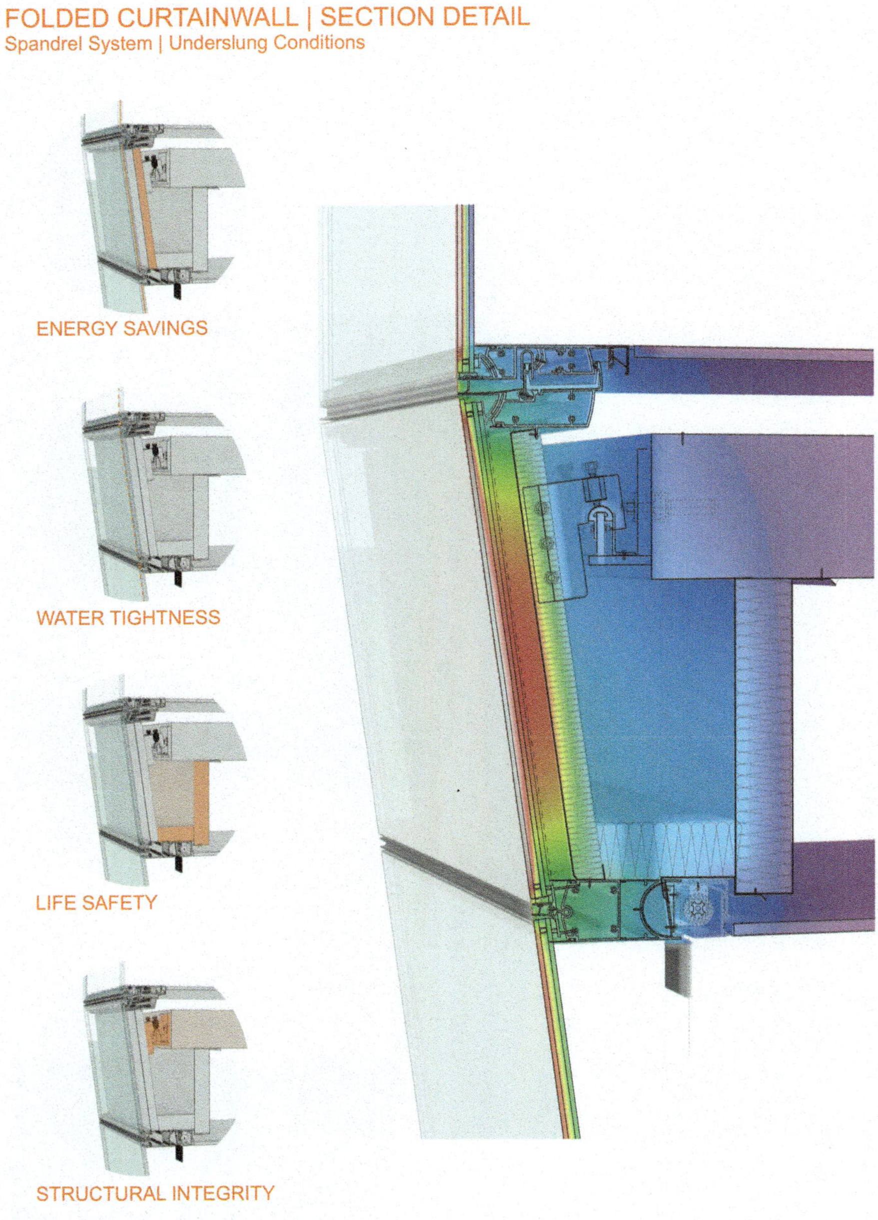

Fig. 4.9 Section detail of the folded curtainwall spandrel at the underslung condition

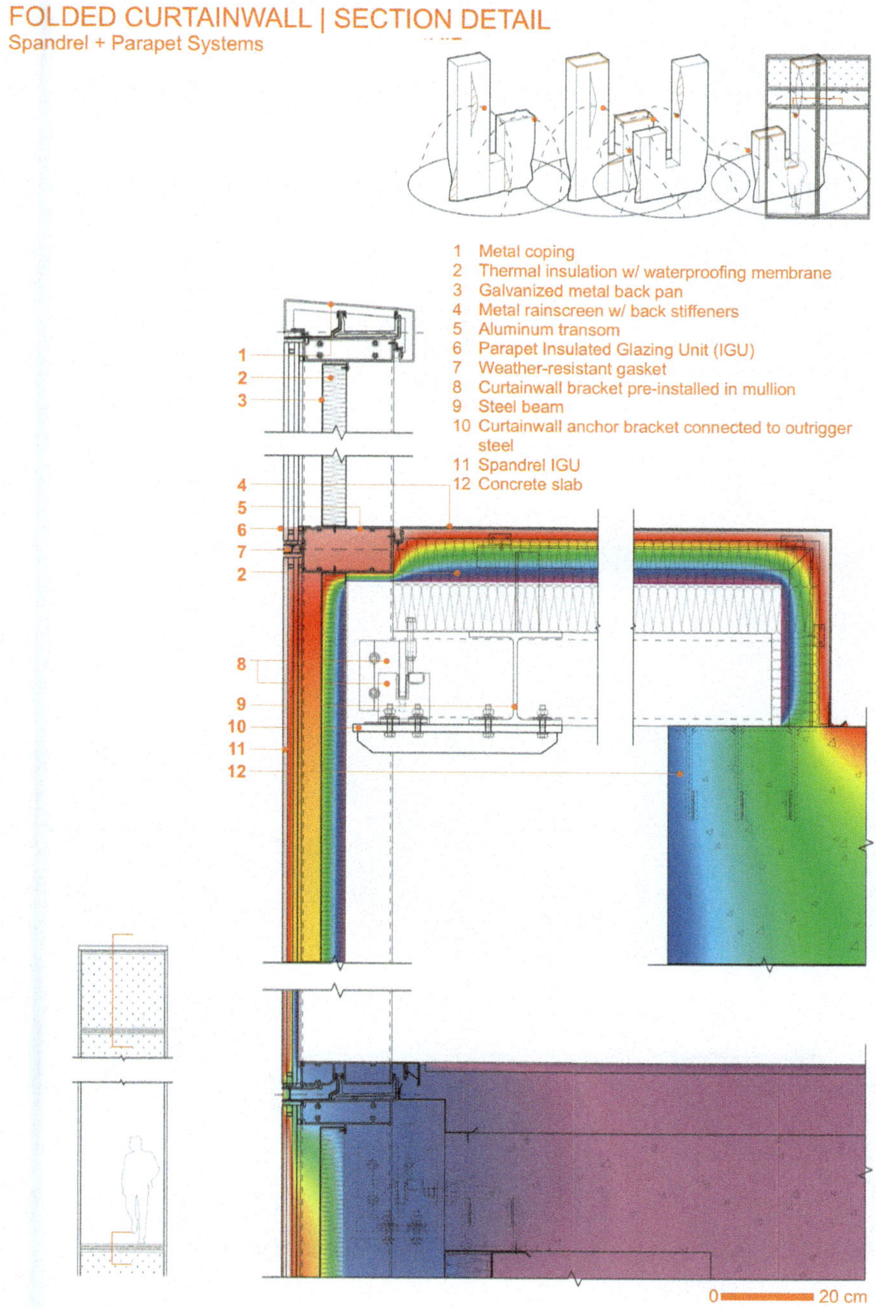

Fig. 4.10 Section detail of the folded curtainwall at the parapet level

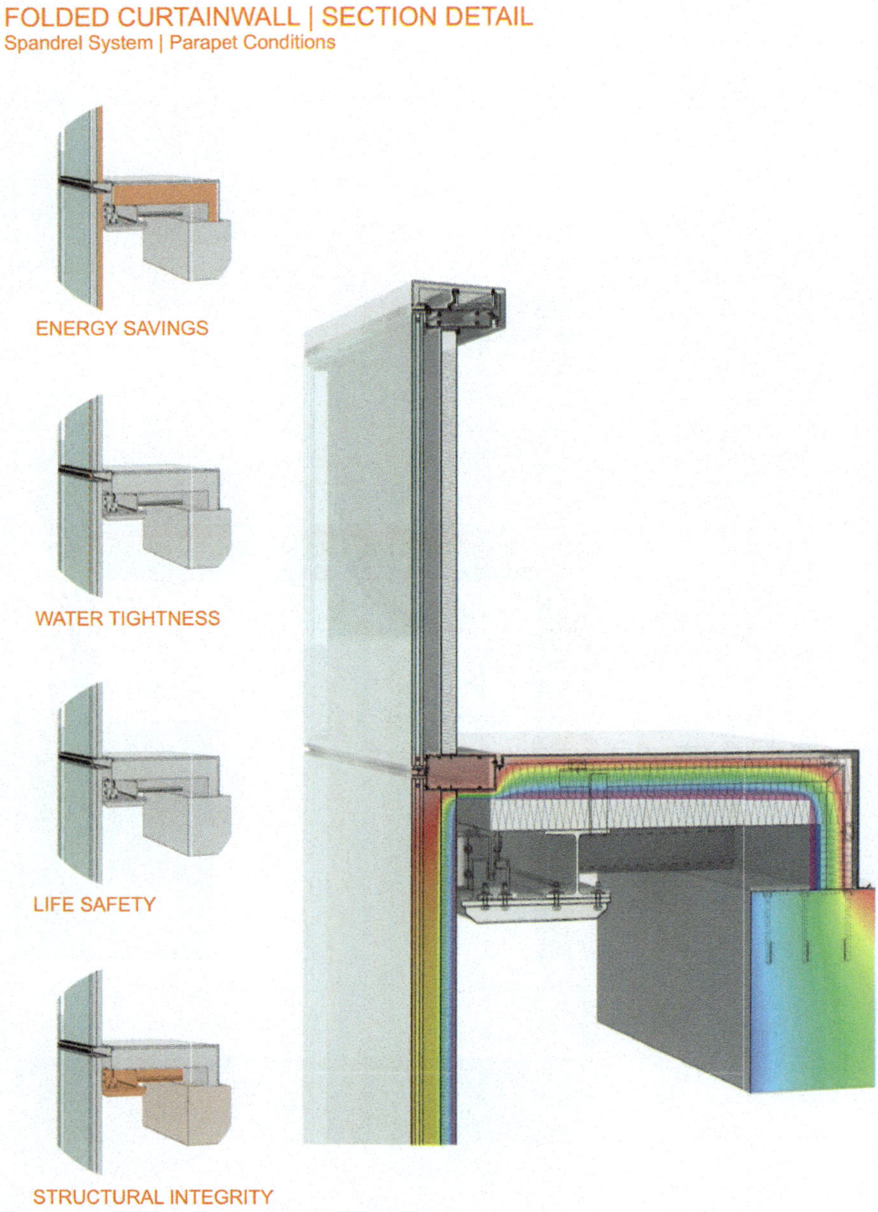

Fig. 4.11 Section detail of the folded curtainwall spandrel panel at the parapet level

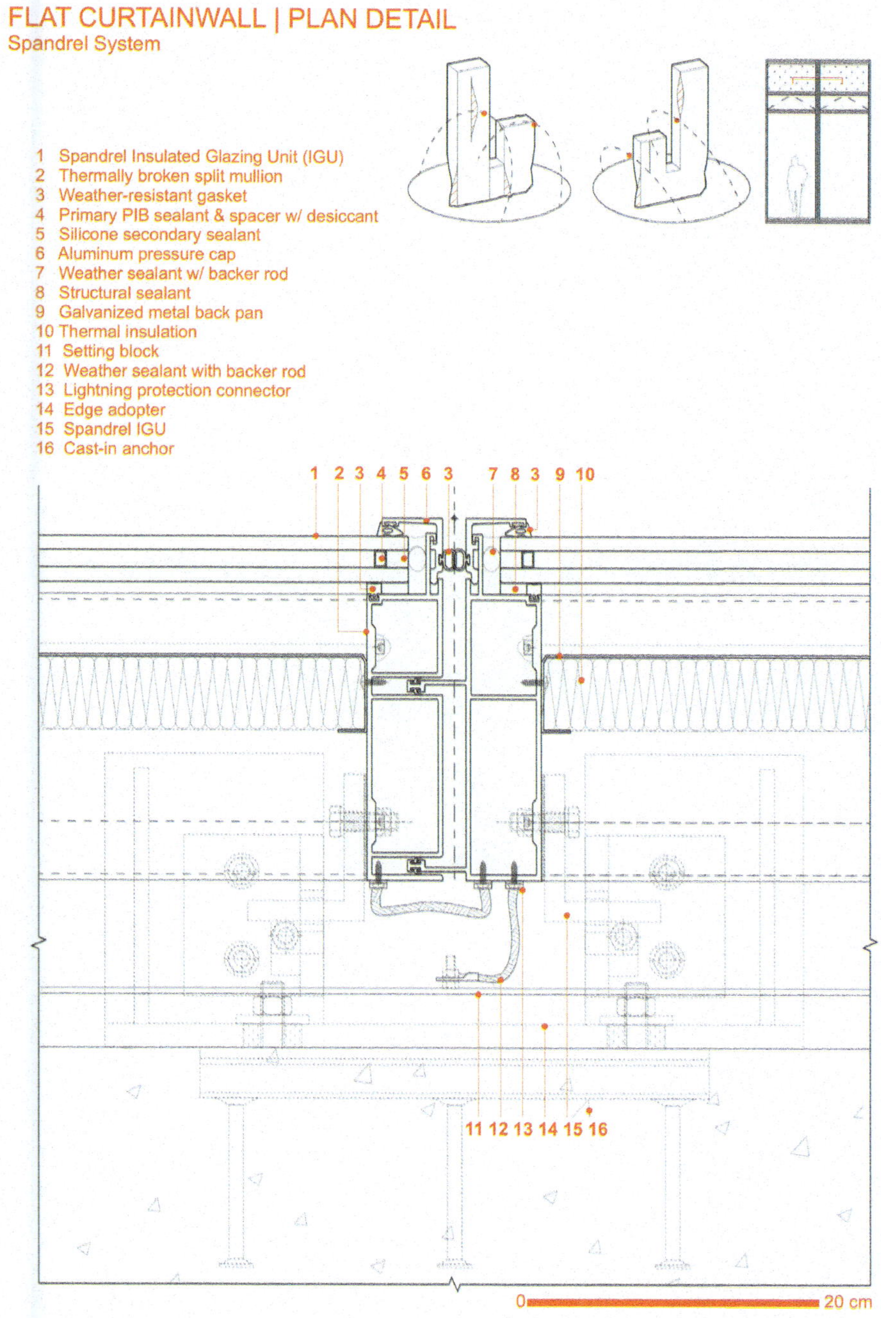

Fig. 4.12 Plan detail of the flat curtainwall spandrel panel

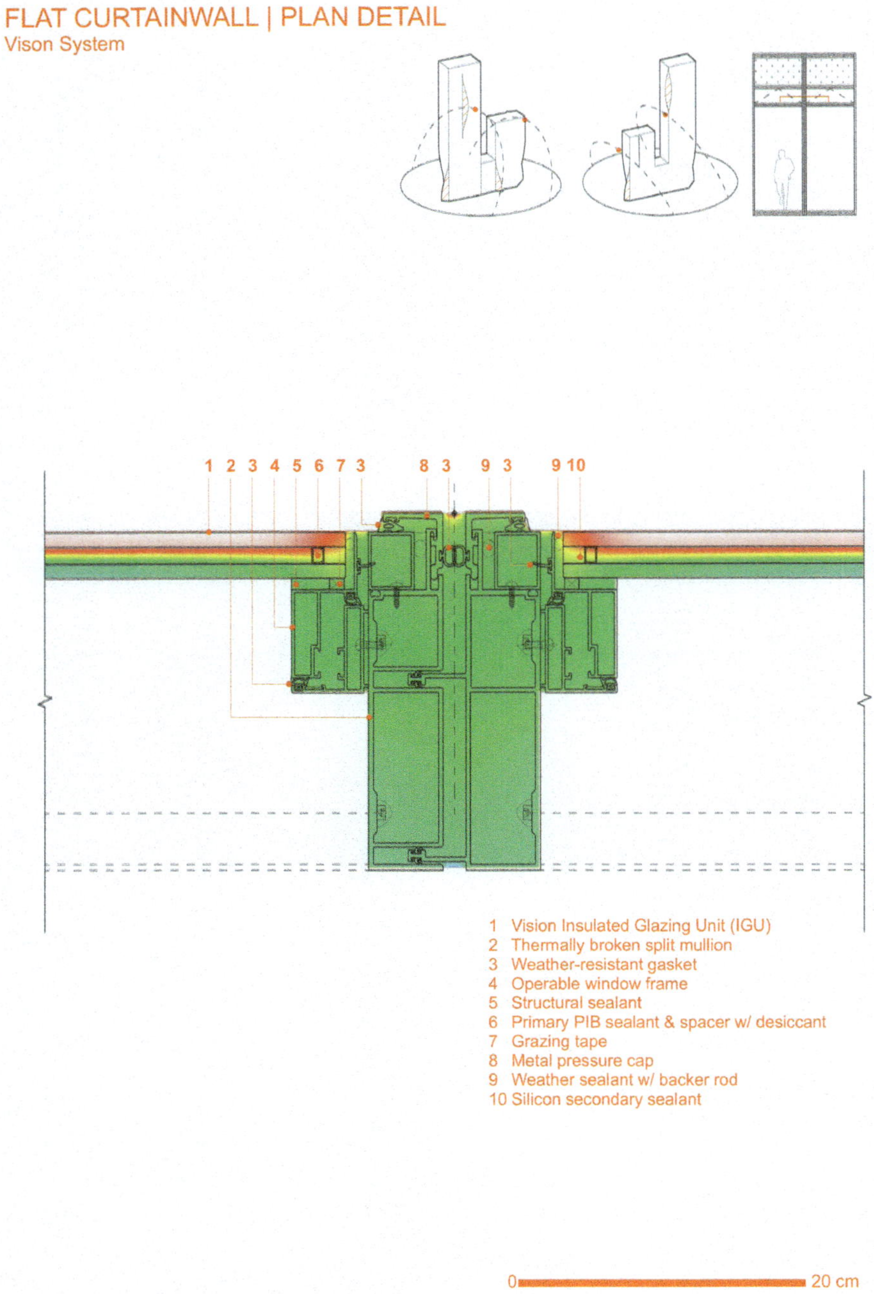

Fig. 4.13 Plan detail of the flat curtainwall vision panel

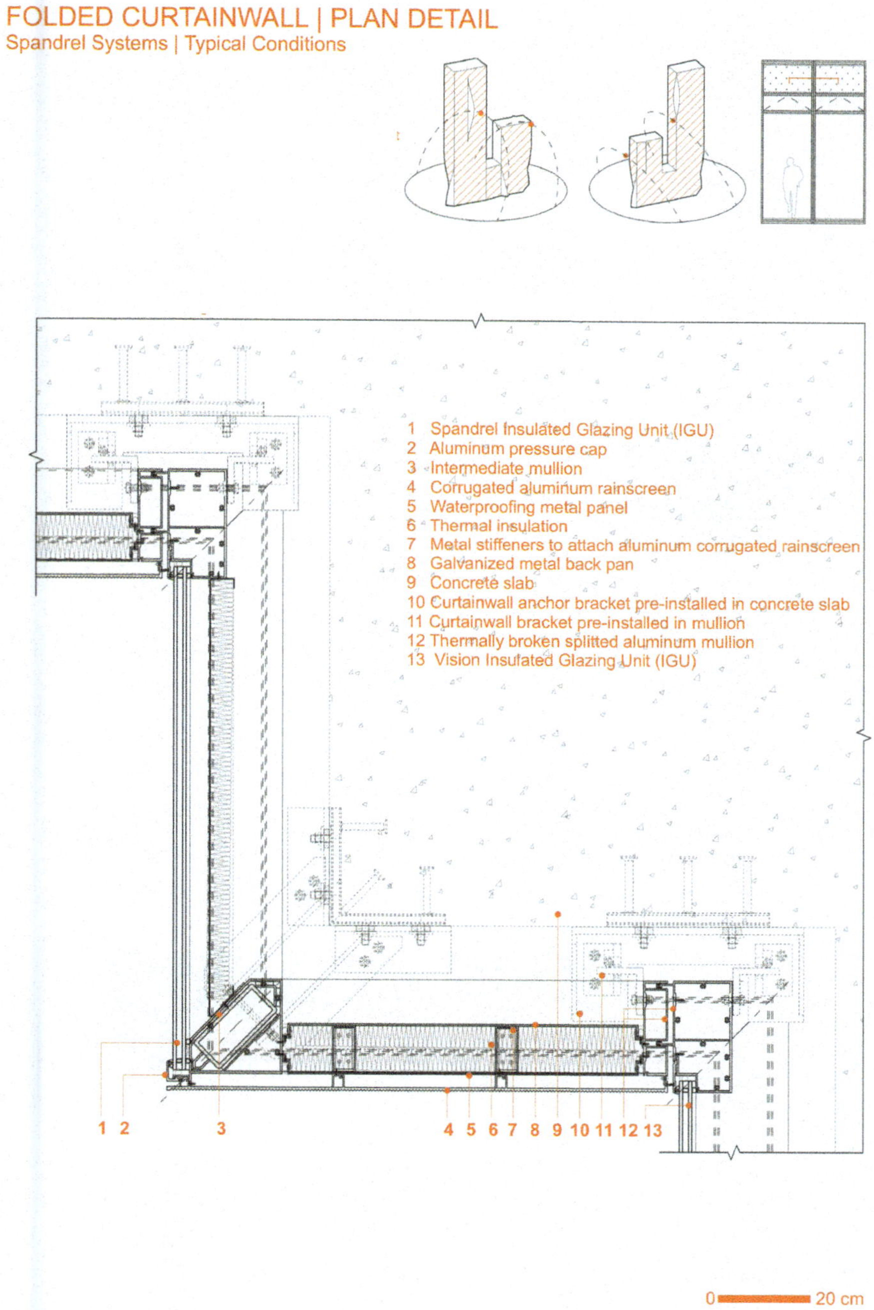

Fig. 4.14 Plan detail of the folded curtainwall spandrel

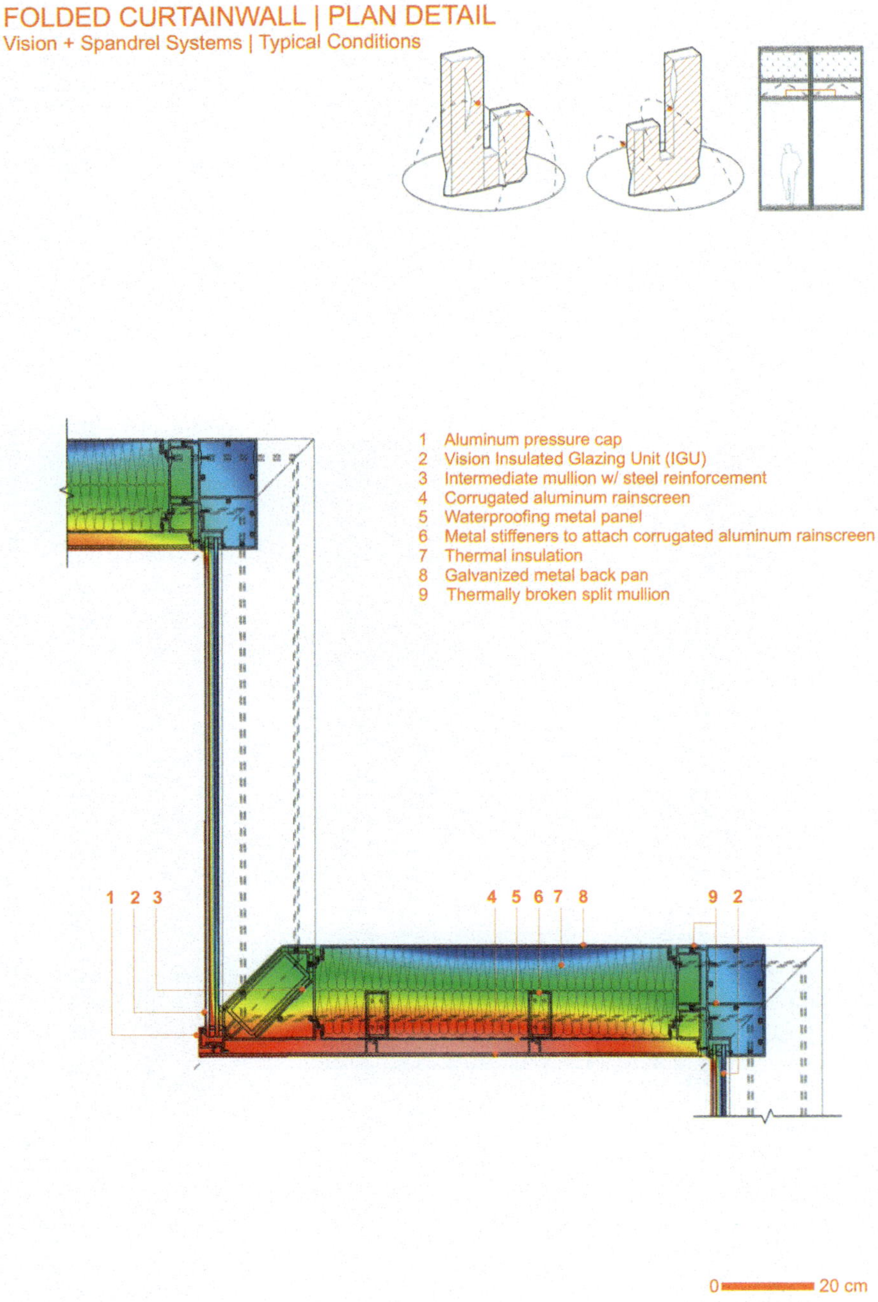

Fig. 4.15 Plan detail of the folded curtainwall vision panel

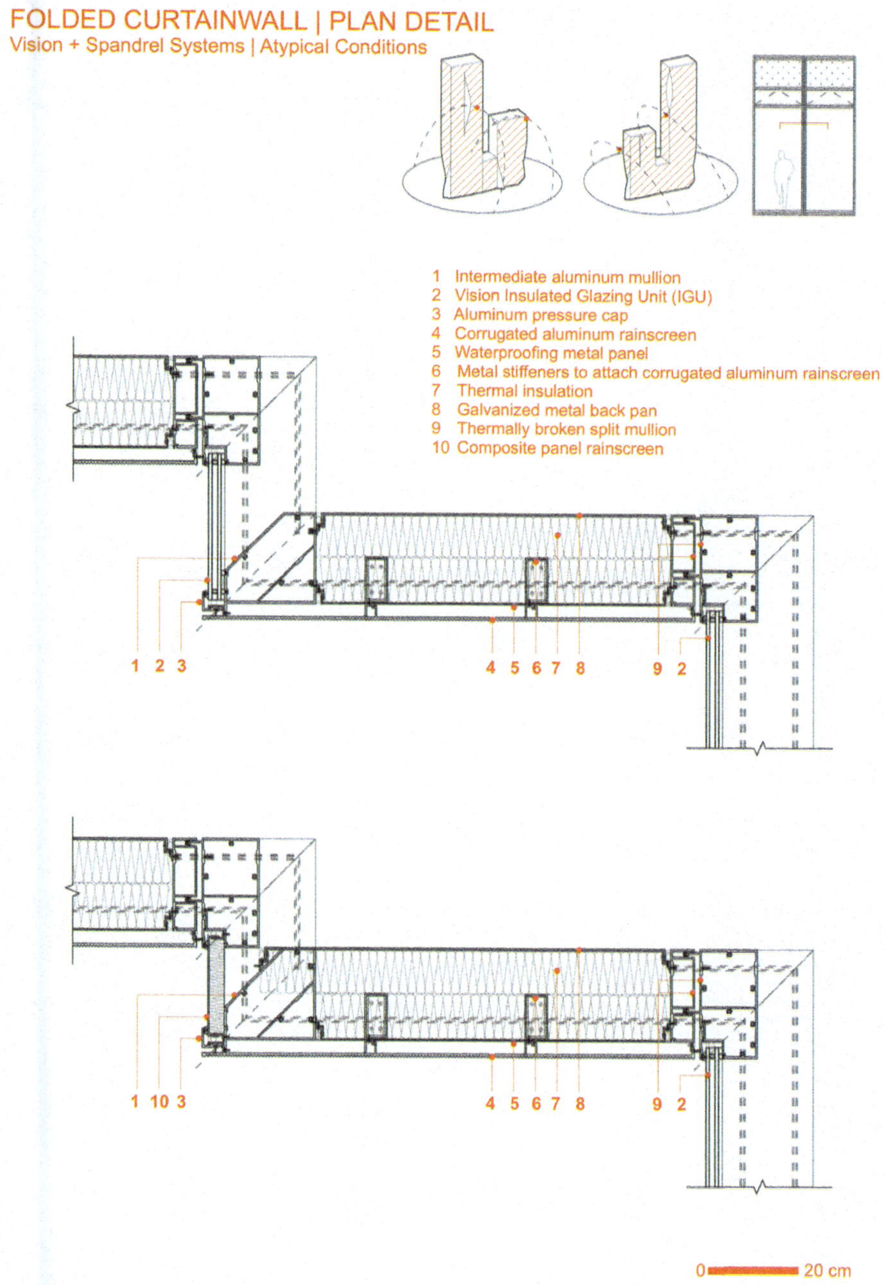

Fig. 4.16 Atypical plan details of the folded curtainwall

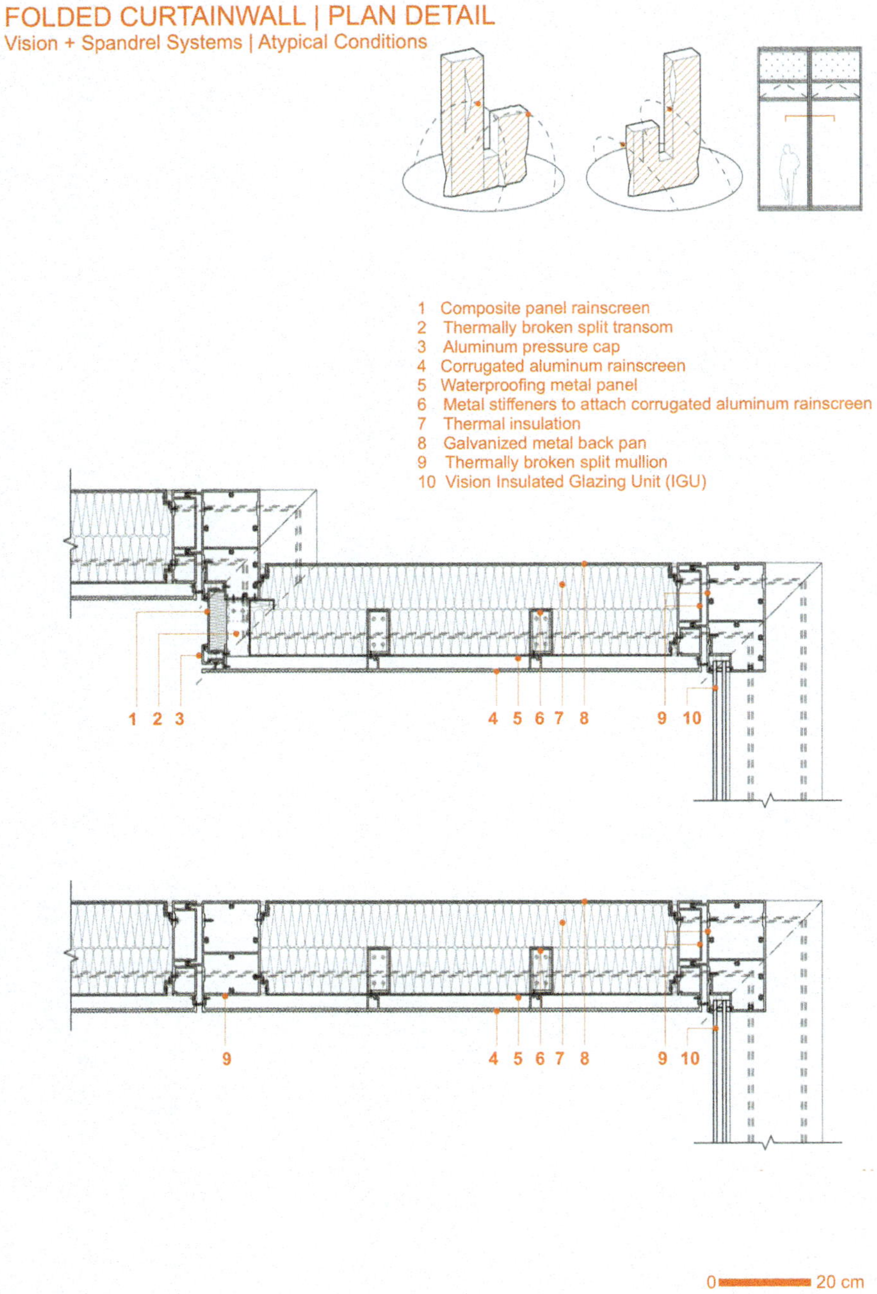

Fig. 4.17 Atypical plan detail of the folded curtainwall

FOLDED CURTAINWALL | PLAN DETAIL

Spandrel Systems | Corner Conditions

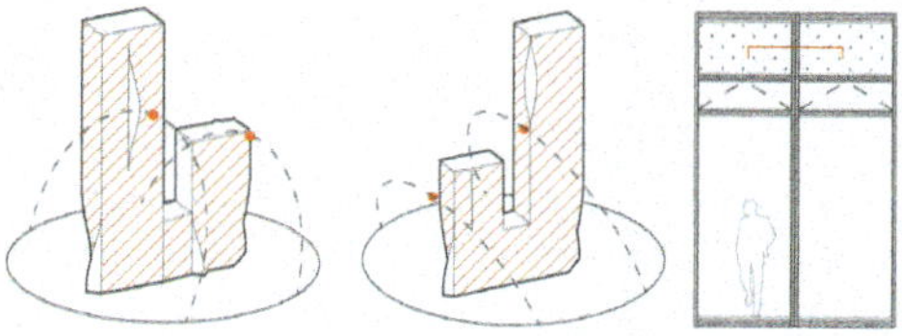

1 Spandrel Insulated Glazing Unit (IGU)
2 Aluminum pressure cap
3 Intermediate mullion
4 Corrugated aluminum rainscreen
5 Waterproofing metal panel
6 Thermal insulation
7 Metal stiffeners to attach aluminum rainscreen
8 Galvanized metal back pan
9 Curtainwall bracket pre-installed in mullion
10 Curtainwall anchor bracket pre-installed in concrete slab
11 Concrete slab
12 Thermally broken split mullion

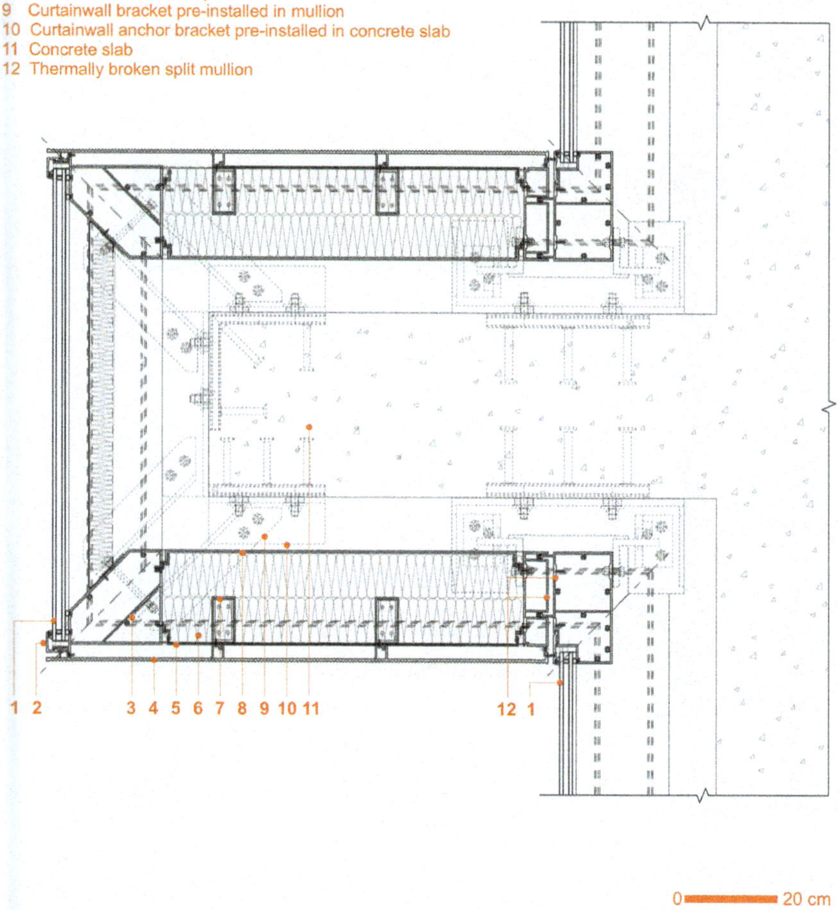

Fig. 4.18 Plan detail of the folded curtainwall spandrel at the corner condition

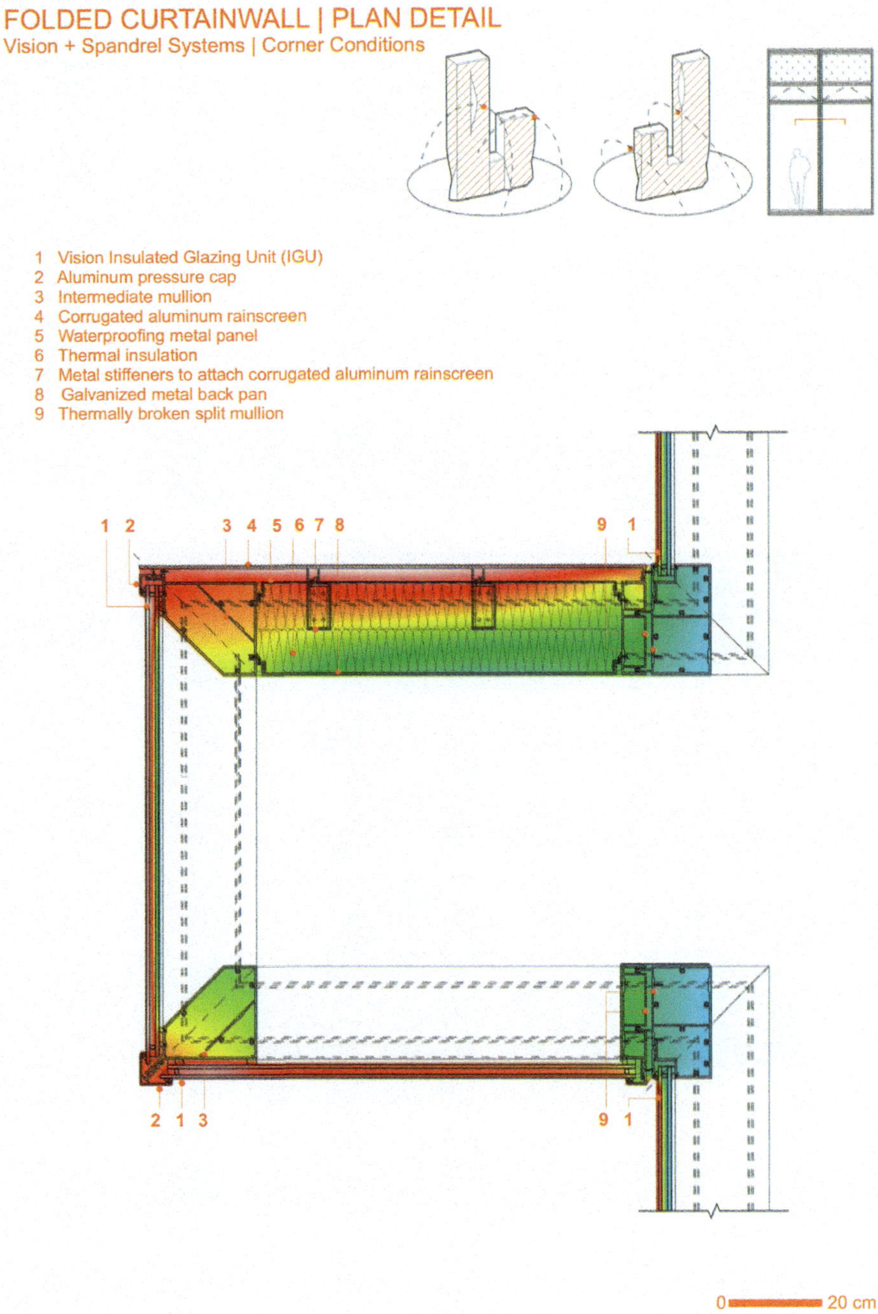

Fig. 4.19 Plan detail of the folded curtainwall vision panel at the corner condition

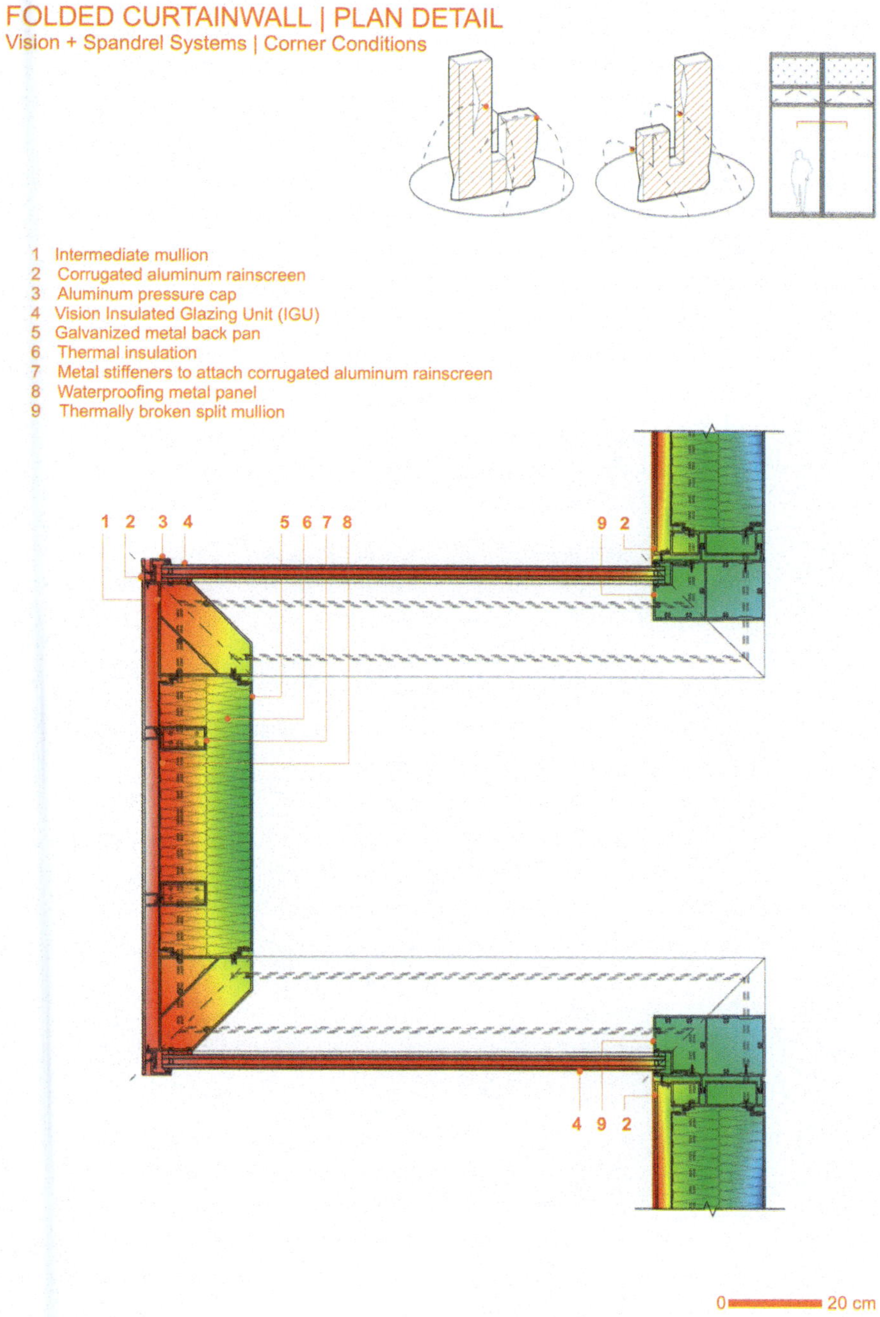

Fig. 4.20 Plan detail of the folded curtainwall vision panel at the corner condition

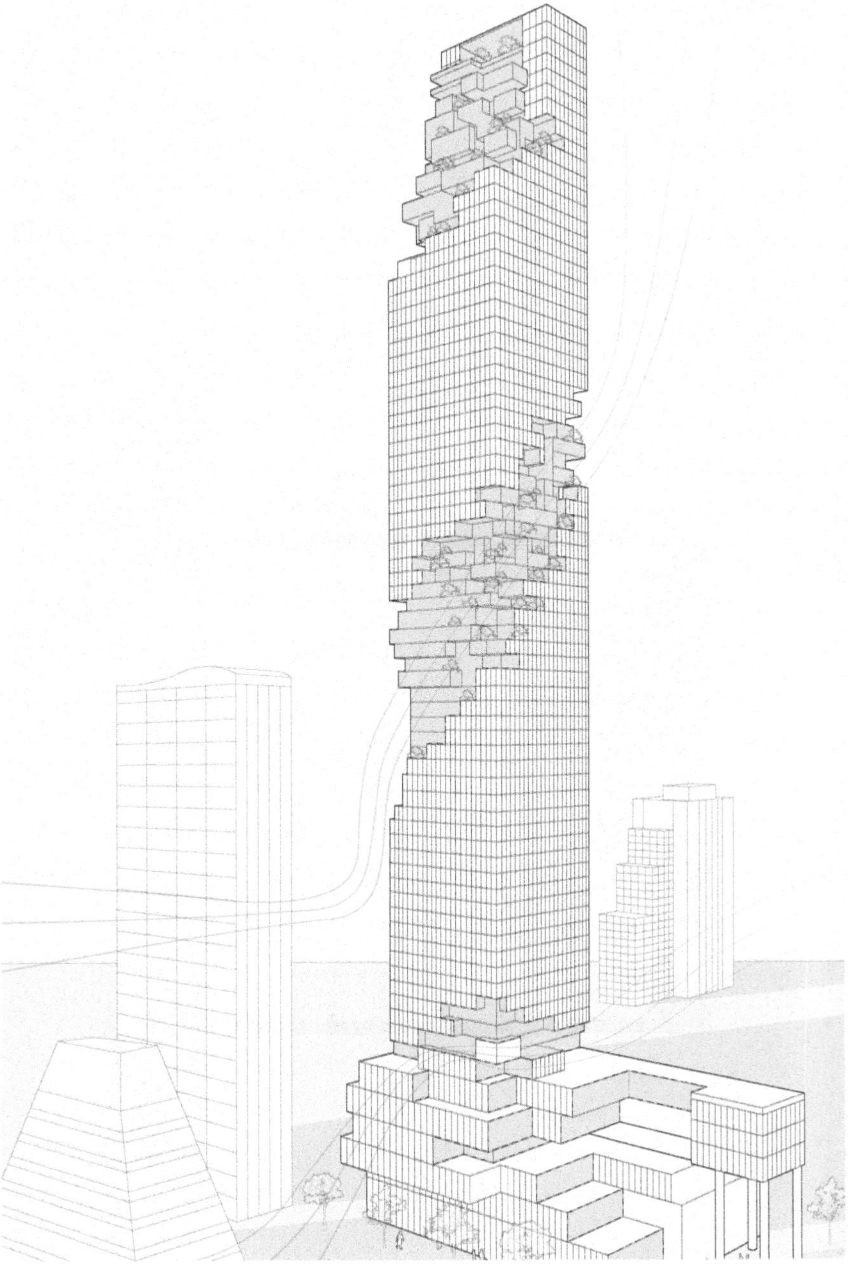

Fig. 4.21 Exterior view of MahaNakhon, Bangkok, Thailand (drawings reproduced from Buro Ole Sheeren)

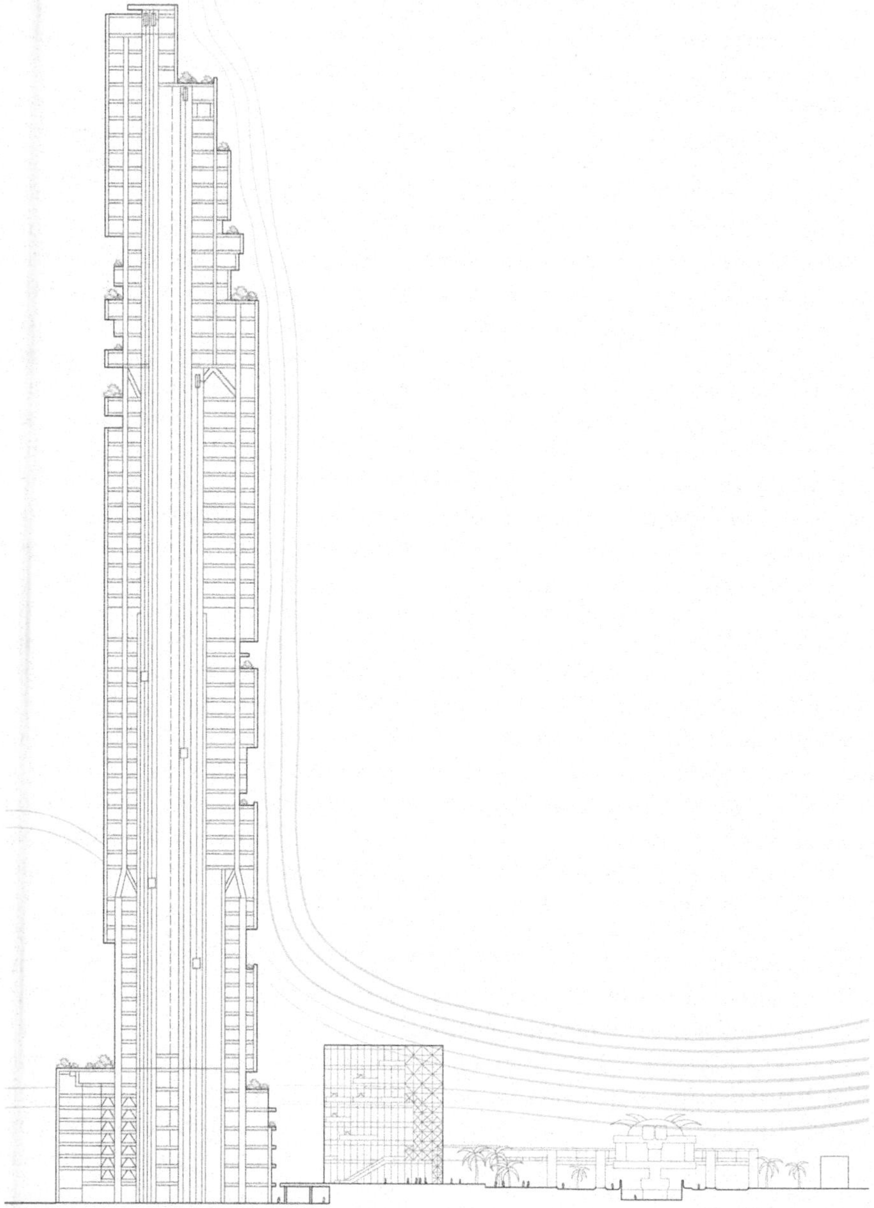

Fig. 4.22 Building section view of MahaNakhon, Bangkok, Thailand (drawings reproduced from Buro Ole Sheeren)

primary structures combine a steel frame with post-tensioned concrete, improving seismic performance, structural efficiency, and space utilization. This innovative approach has the potential for global application, allowing architects to create taller, more resilient skyscrapers that maintain a balance between aesthetics, functionality, safety, and sustainability even in contexts with limited resources [278].

Impact of Tall Towers in Bangkok: Tall buildings can maximize land-use efficiency, potentially preserve green space, reduce urban sprawl, and support connected public transit networks. They create impressive skylines that attract tourists and investors, boosting the city's economy [279]. These structures provide more space for businesses and residents while minimizing the overall infrastructure footprint. Tall buildings are increasingly common in Bangkok's skyline. Their rapid construction necessitates careful consideration of environmental impacts. Oldfield (2019) offers insights into designing these structures sustainably, addressing both environmental benefits and material-related concerns aimed at reducing energy use and waste production, while also considering other factors such as air quality and water conservation [280]. He further suggests that materials used in the construction of tall towers must not only meet safety requirements but also maintain long-term durability. Additionally, he outlines strategies for mitigating the potential impacts of urban heat islands, created by large concentrations of concrete surfaces, which can help reduce temperatures during hot summer months when air conditioning is heavily relied upon throughout Bangkok's metropolitan area.

As of 2024, the population of Bangkok, the capital city of Thailand, is approximately 11 million spread over an area of approximately 1569 km^2 (605.8 mi^2), and it is among the largest cities in Southeast Asia, characterized by a compact urban core and extended suburbs [281]. It is situated on the delta of Chao Phraya River in Central Thailand, which has contributed to its development as a major business hub for regional and international commerce due to its river network and proximity to the Gulf of Thailand. Bangkok's transformation from a canal-based settlement to a modern, land-based metropolis is central to its identity [282]. The city's fluid urban spaces, water-based culture, vibrant informal sector, and prominent religious sites continue to remain key elements of its unique character.

This city has been transforming at a rapid pace over the past few decades into a highly urbanized environment, characterized by tall buildings, shopping centers, and well-developed infrastructure. Traffic congestion, flooding, and pollution are some of the issues that have compelled authorities to improve transportation networks, such as BTS Skytrain and the MRT subway, to achieve better urban connectivity [283]. Bangkok experiences a tropical savanna climate (Köppen: Aw) [284]. A clear distinction exists between the wet and dry season, with high temperatures throughout the year and an annual rainfall of 1207 mm (47.5 in). The rainy season spans from May to October, during which monsoonal rains are heavy, while November throughout to April is typically the dry season (Figs. 4.23, 4.24, 4.25, 4.26, 4.27, 4.28, 4.29, 4.30, 4.31, 4.32, 4.33, 4.34, 4.35, 4.36, 4.37, 4.38, 4.39, 4.40 and 4.41).

Fig. 4.23 MahaNakhon, Bangkok, Thailand (2018), and sustainable façade applications

Location	MahaNakhon, Thailand
Completion date	2018
Köppen Climate Classification	tropical savanna climate (Köppen climate: Aw)
Building owner/primary tenant	PACE Development
Architects & design teams	Ateliers Jean Nouvel
# of floors	76
Gross floor areas (m^2)	150,000 m^2

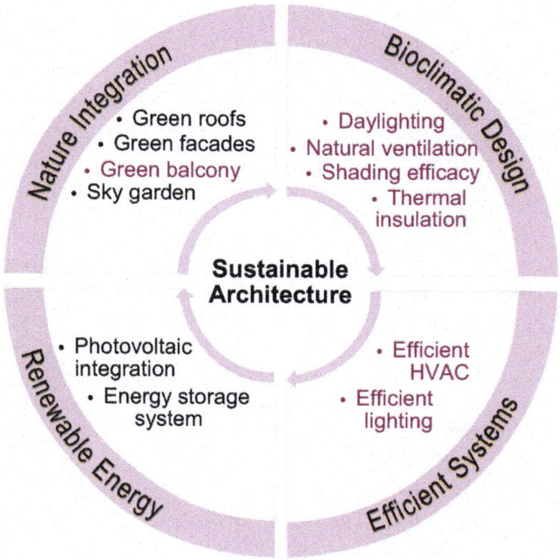

4.3 Le Nouvel KLCC, Kuala Lumpur, Malaysia, 2019

Located in Kuala Lumpur, Malaysia, Le Nouvel KLCC offers two residential towers with 43 and 48 stories and 195 upscale residential units. Jean Nouvel, a French architect who won the Pritzker prize laureate, combines the architectural character of high-rise buildings with functional and ecological design that harmonizes the building with dense urban environments. Nouvel partnered with Patrick Blanc, a botanist known for pioneering vertical vegetation systems, and implemented the living wall façades with extensive flora richness that are rarely matched by comparable developments worldwide [285]. The green wall and vertical garden comprise approximately 820 vertical planter units and over 200 plant species [286]. These living wall systems offer both functional benefits, such as solar shading, natural ventilation, air purification, and enhanced daylighting penetration. The addition of greenery façades improves residents' experience through increased proximity to nature within an urban context,

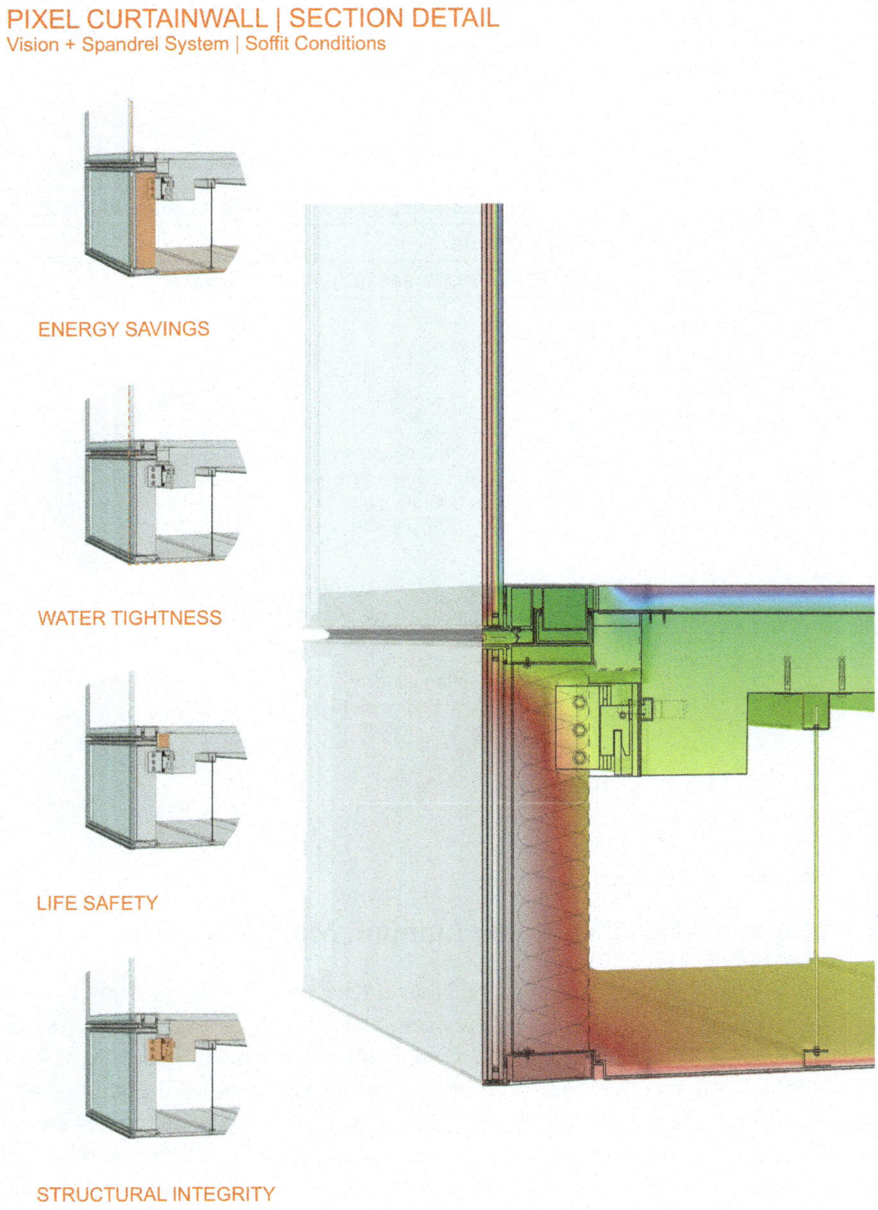

Fig. 4.24 Section detail of the pixel curtainwall at the soffit level

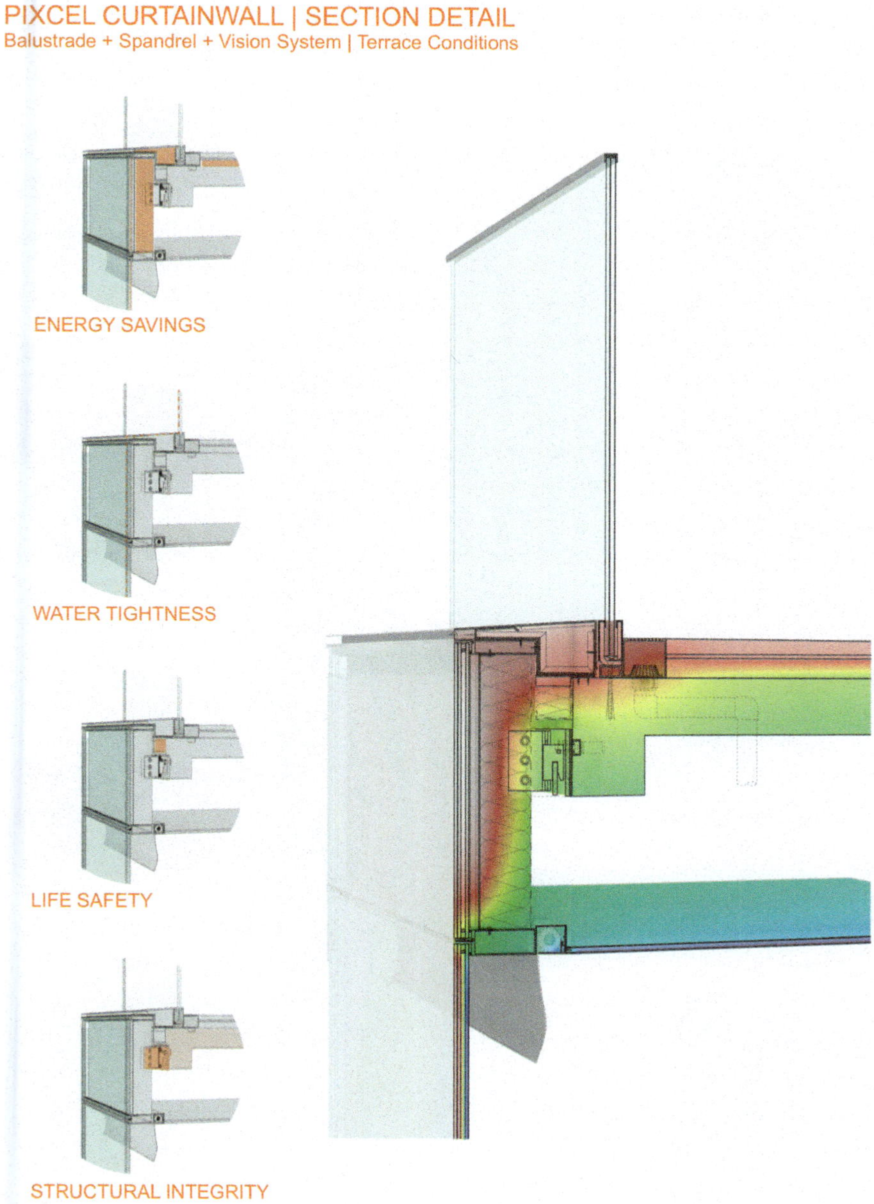

Fig. 4.25 Section detail of the pixel curtainwall at the terrace level

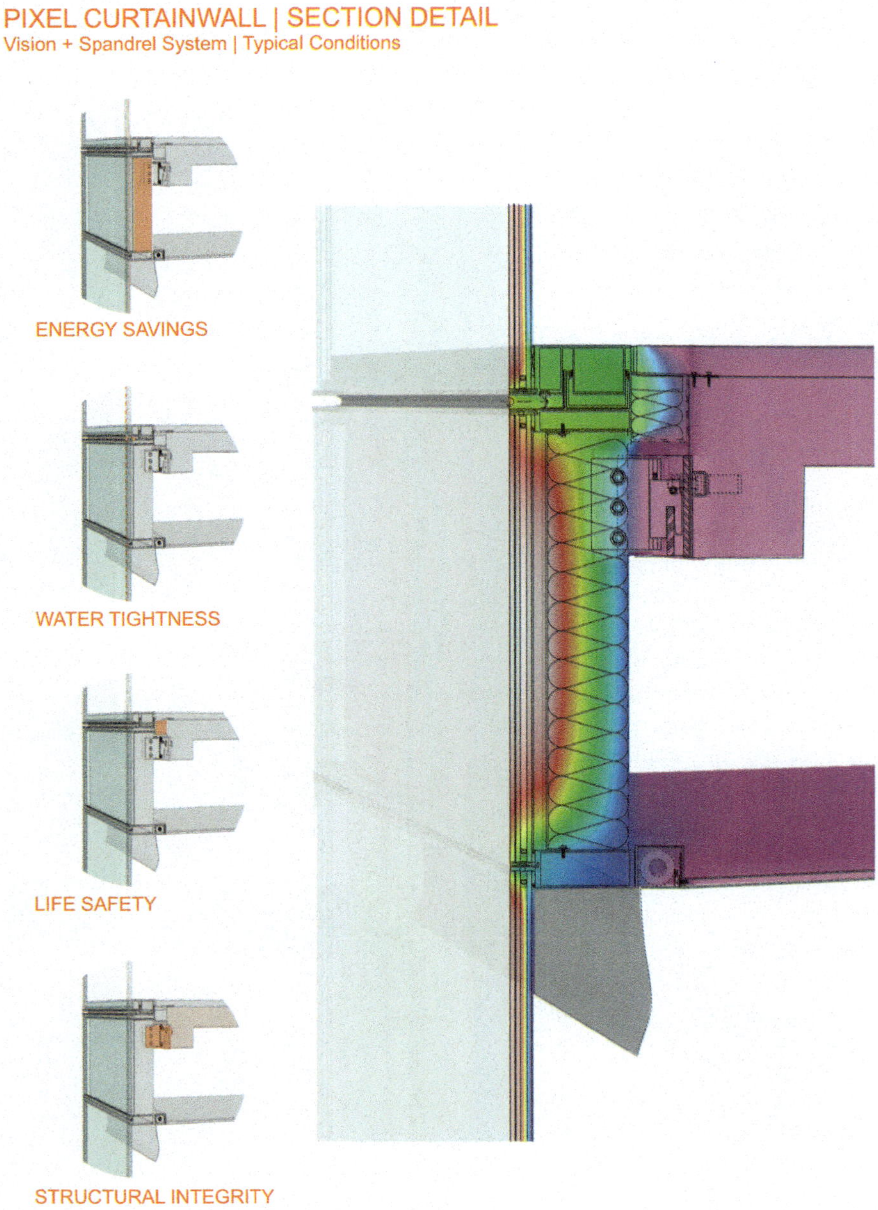

Fig. 4.26 Section detail of the pixel curtainwall

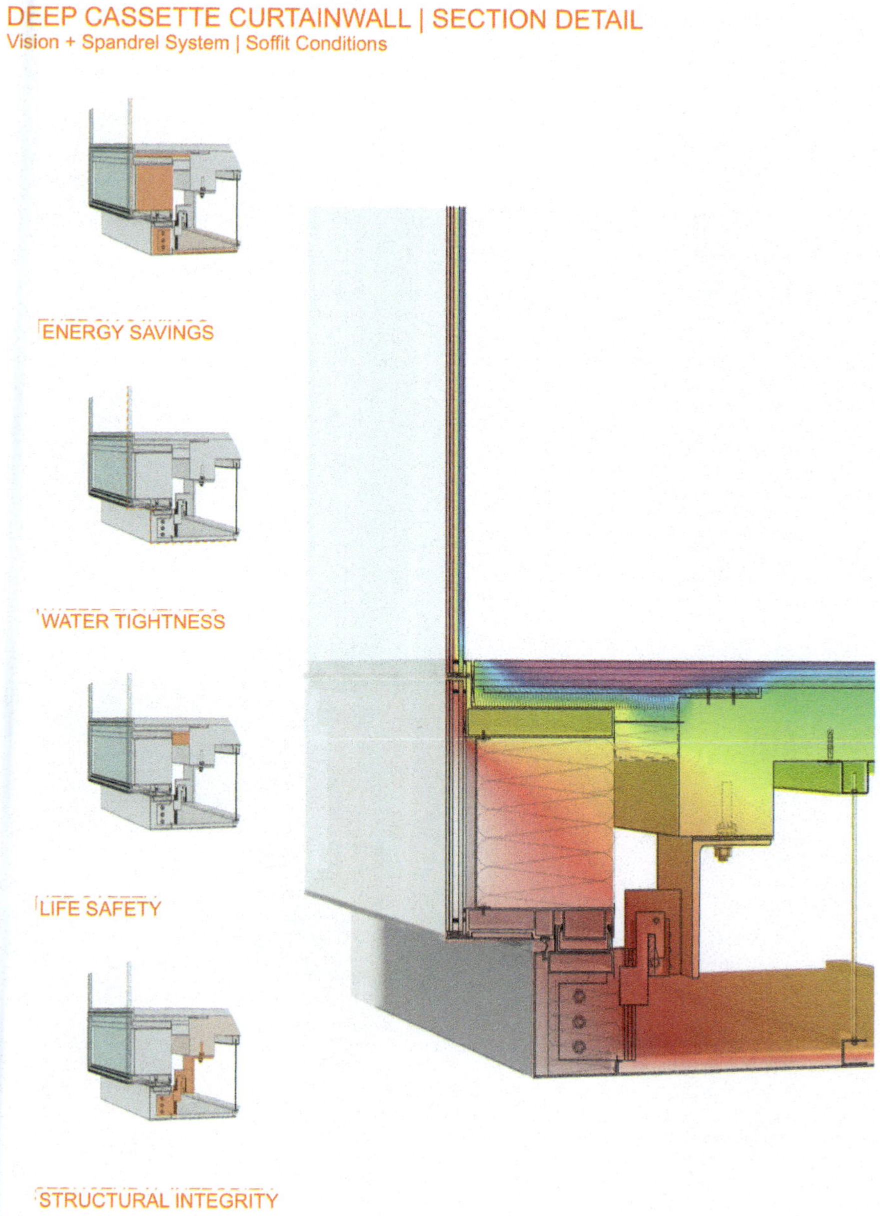

Fig. 4.27 Section detail of the deep cassette curtainwall at the soffit level

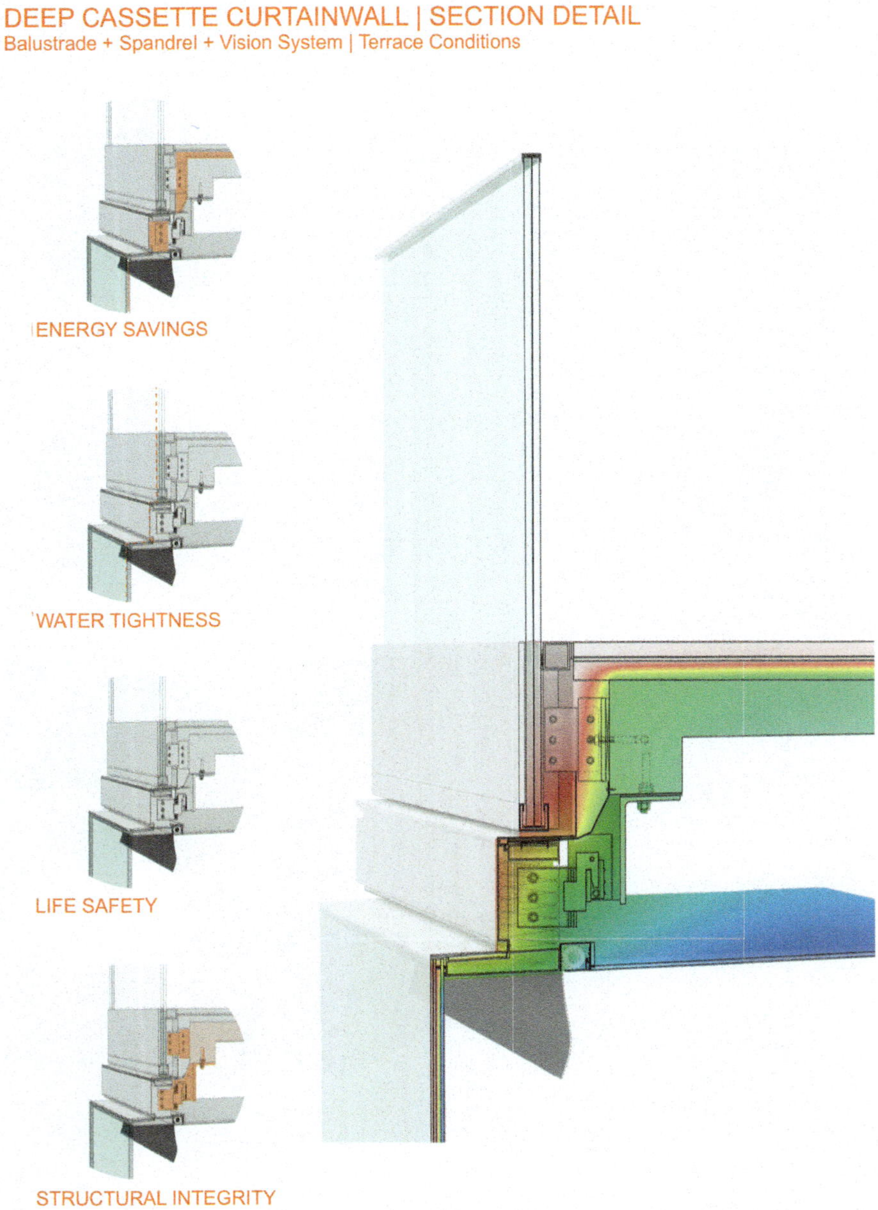

Fig. 4.28 Section detail of the deep cassette curtainwall at the terrace level

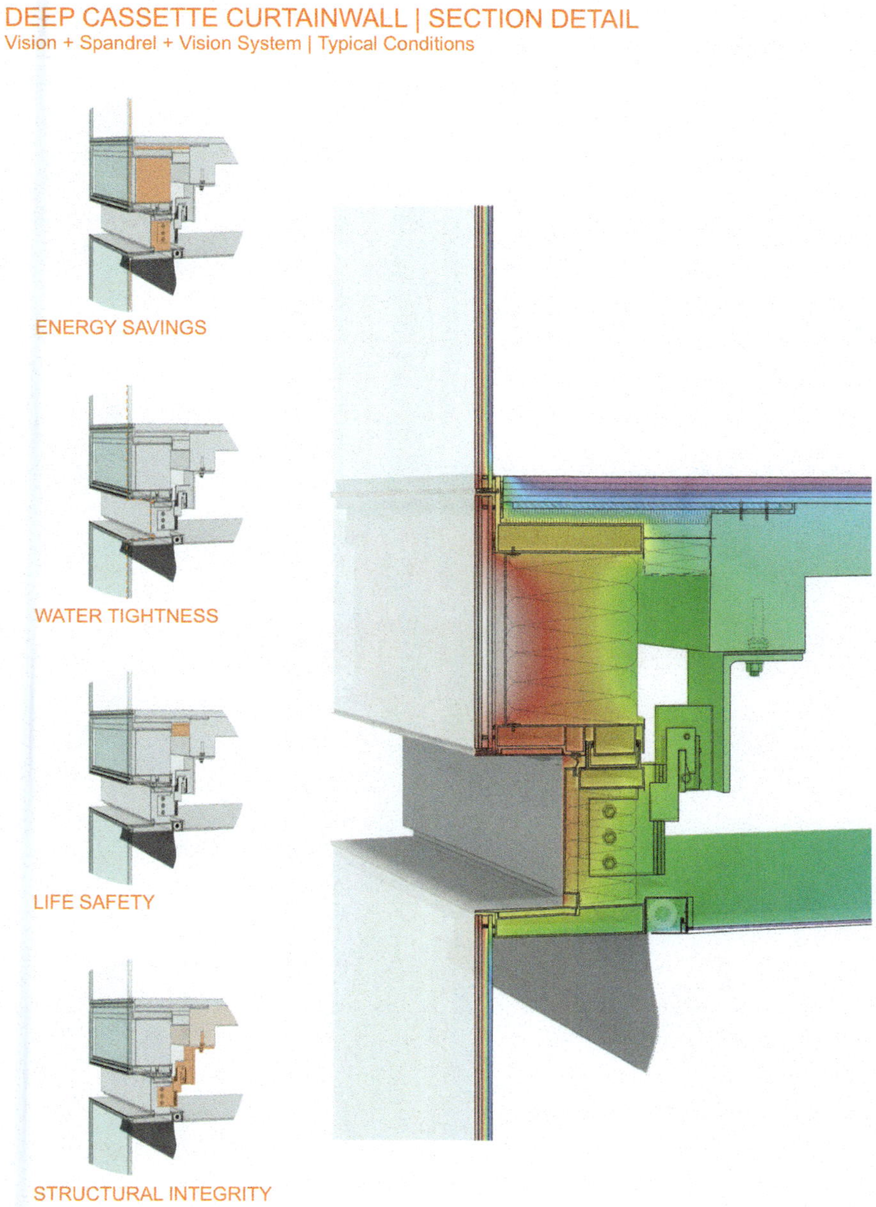

Fig. 4.29 Typical section detail of a deep cassette curtainwall

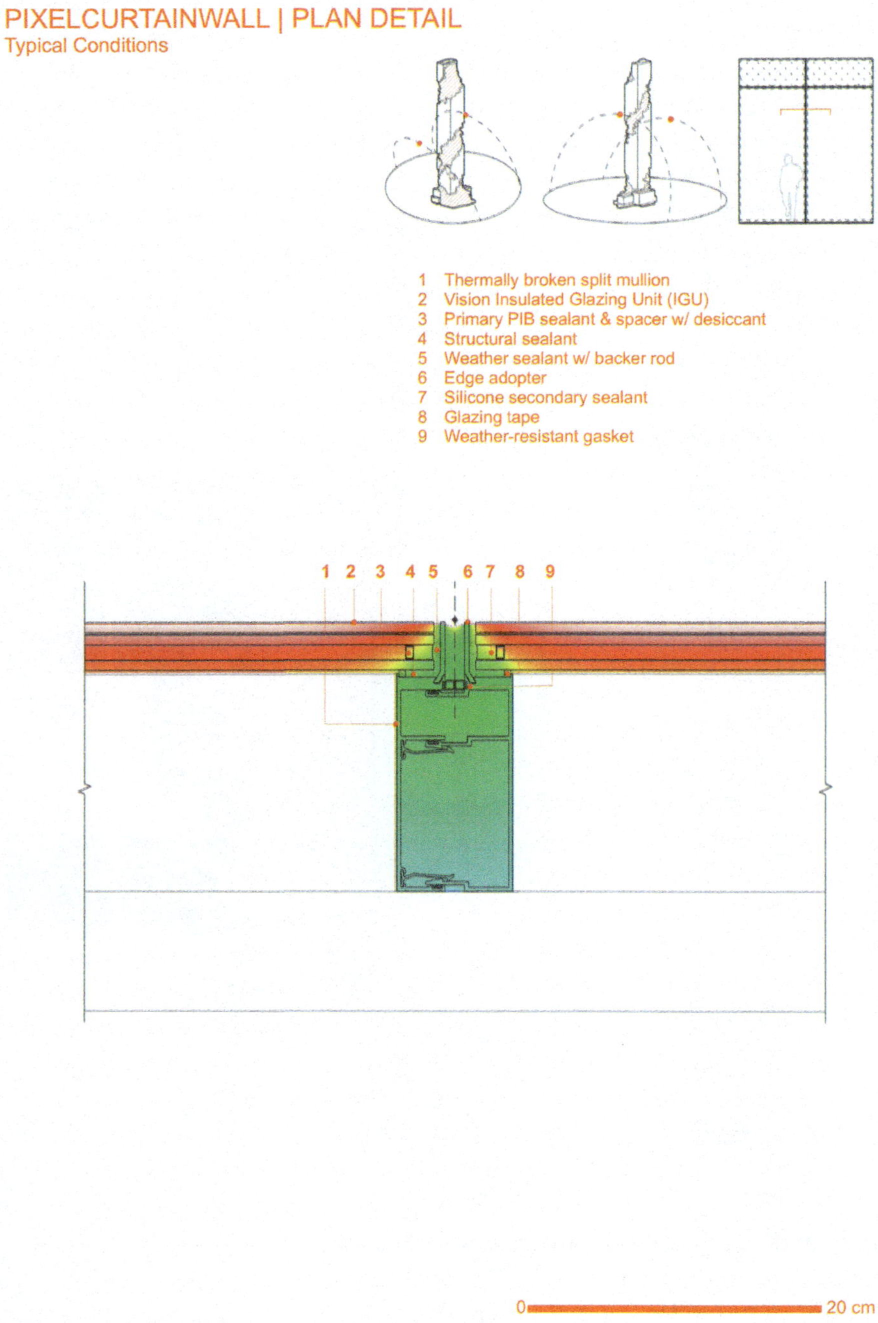

Fig. 4.30 Typical plan detail of the pixel curtainwall

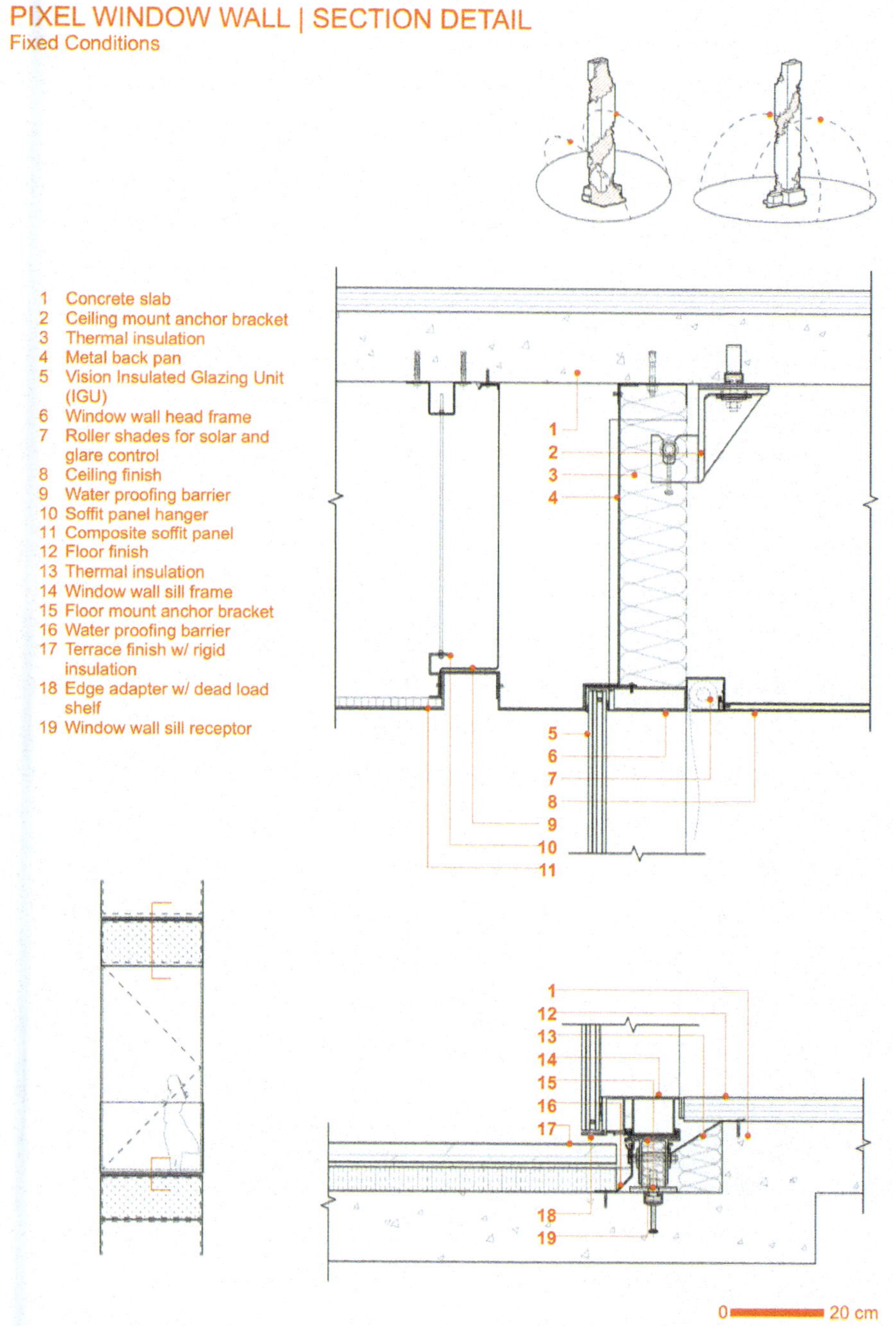

Fig. 4.31 Section detail of the pixel window system

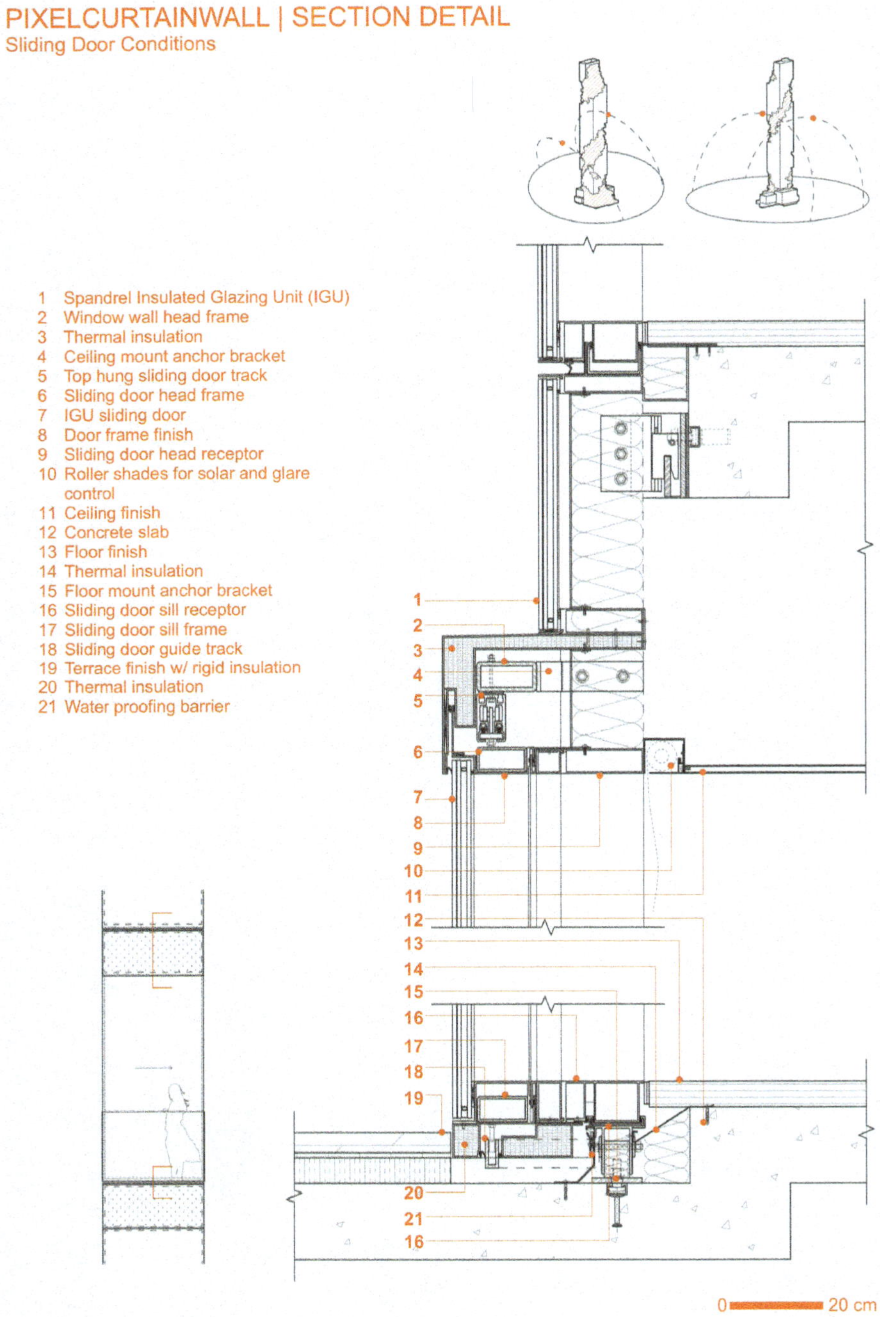

Fig. 4.32 Section detail of the pixel curtainwall with a sliding door

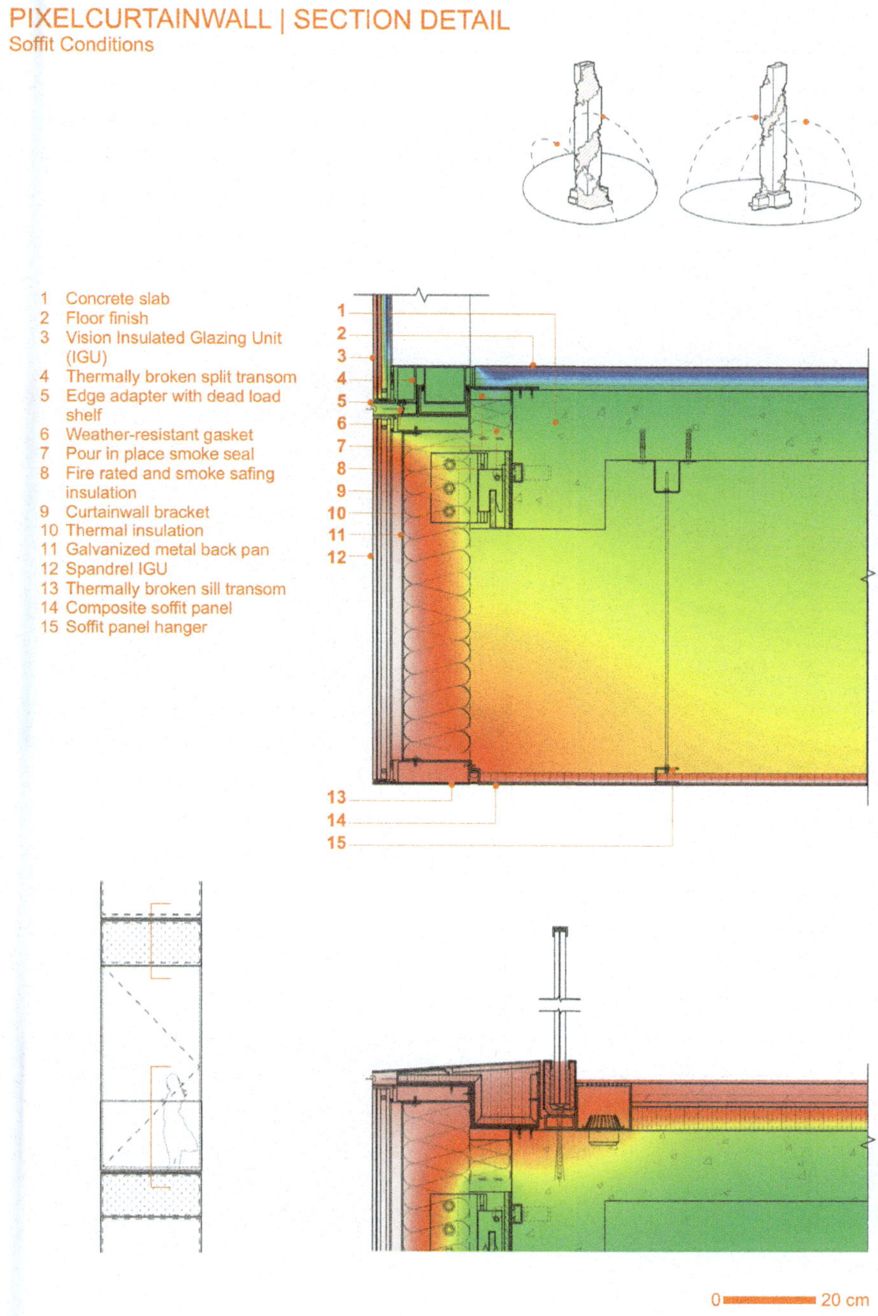

Fig. 4.33 Section detail of the pixel curtainwall at the soffit level

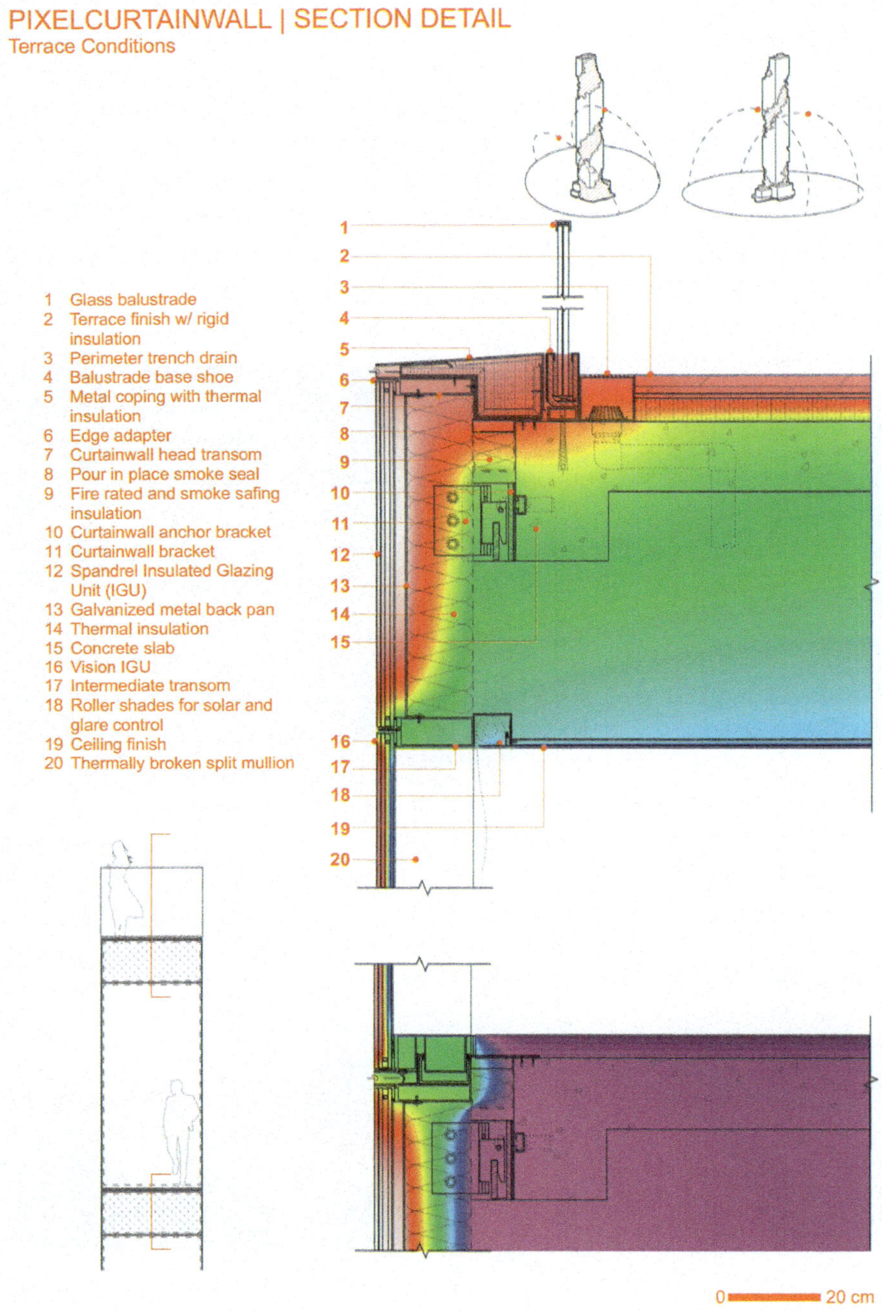

Fig. 4.34 Section detail of the pixel curtainwall at the terrace level

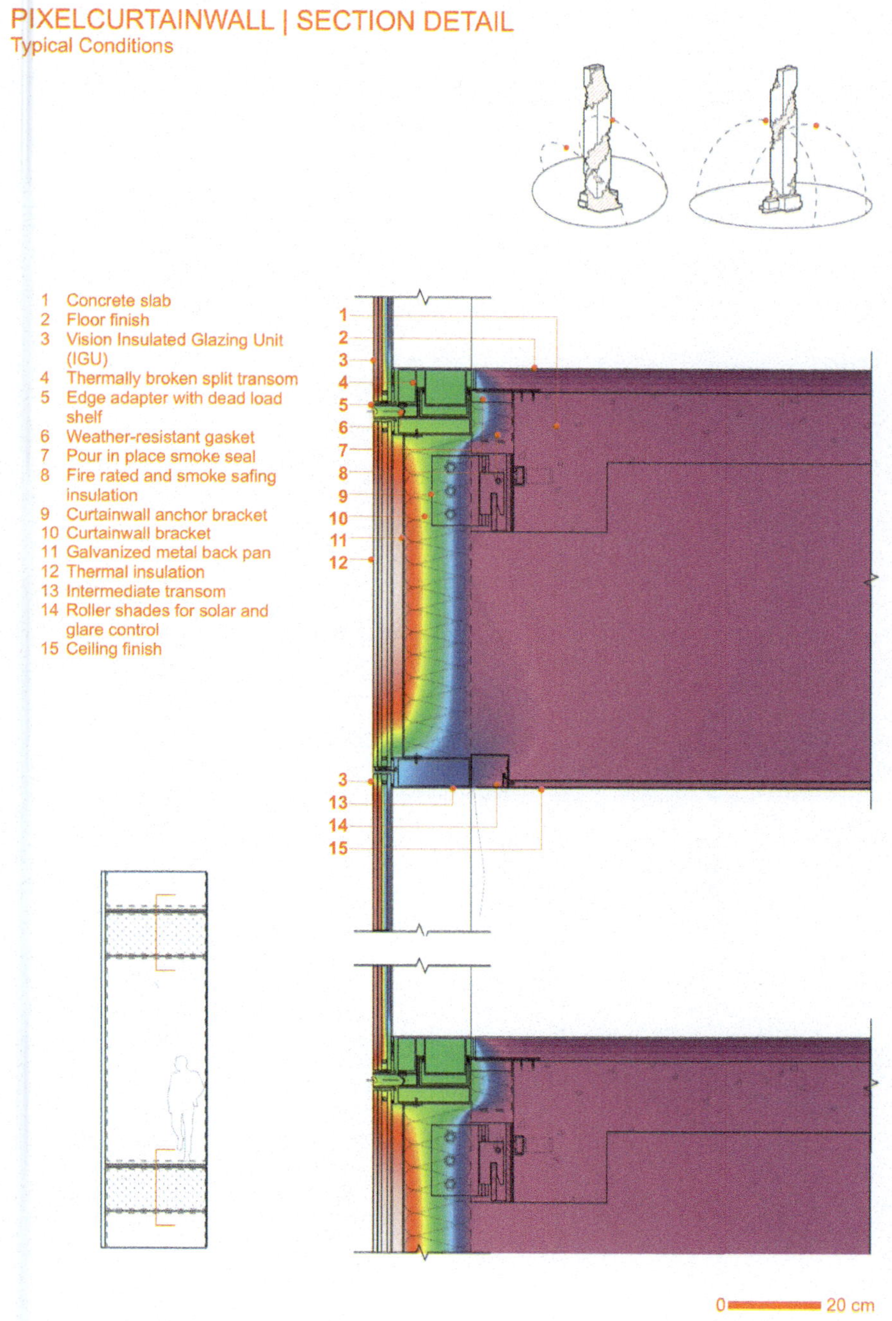

Fig. 4.35 Section detail of the pixel curtainwall

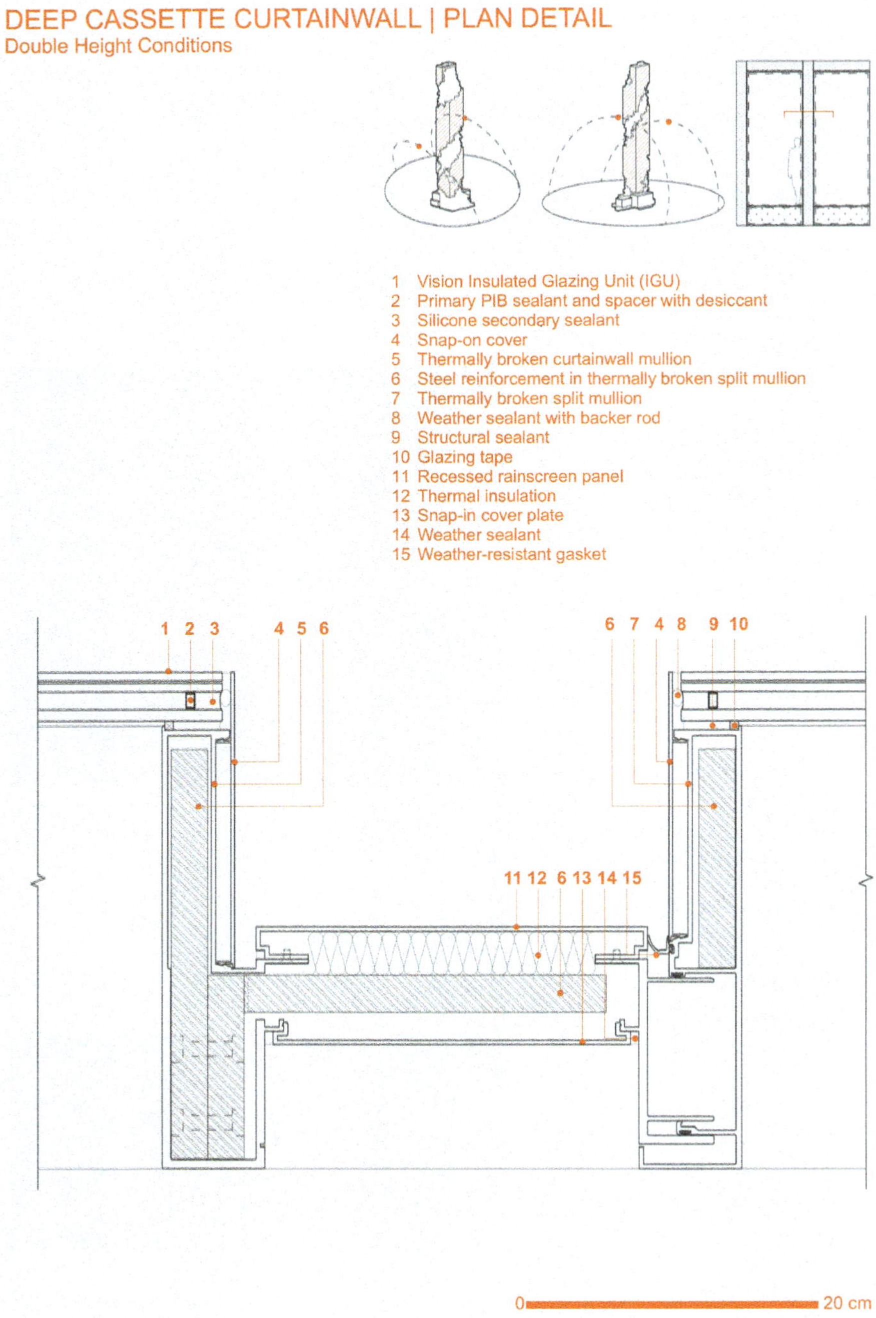

Fig. 4.36 Plan detail of the deep cassette curtainwall at the double height space

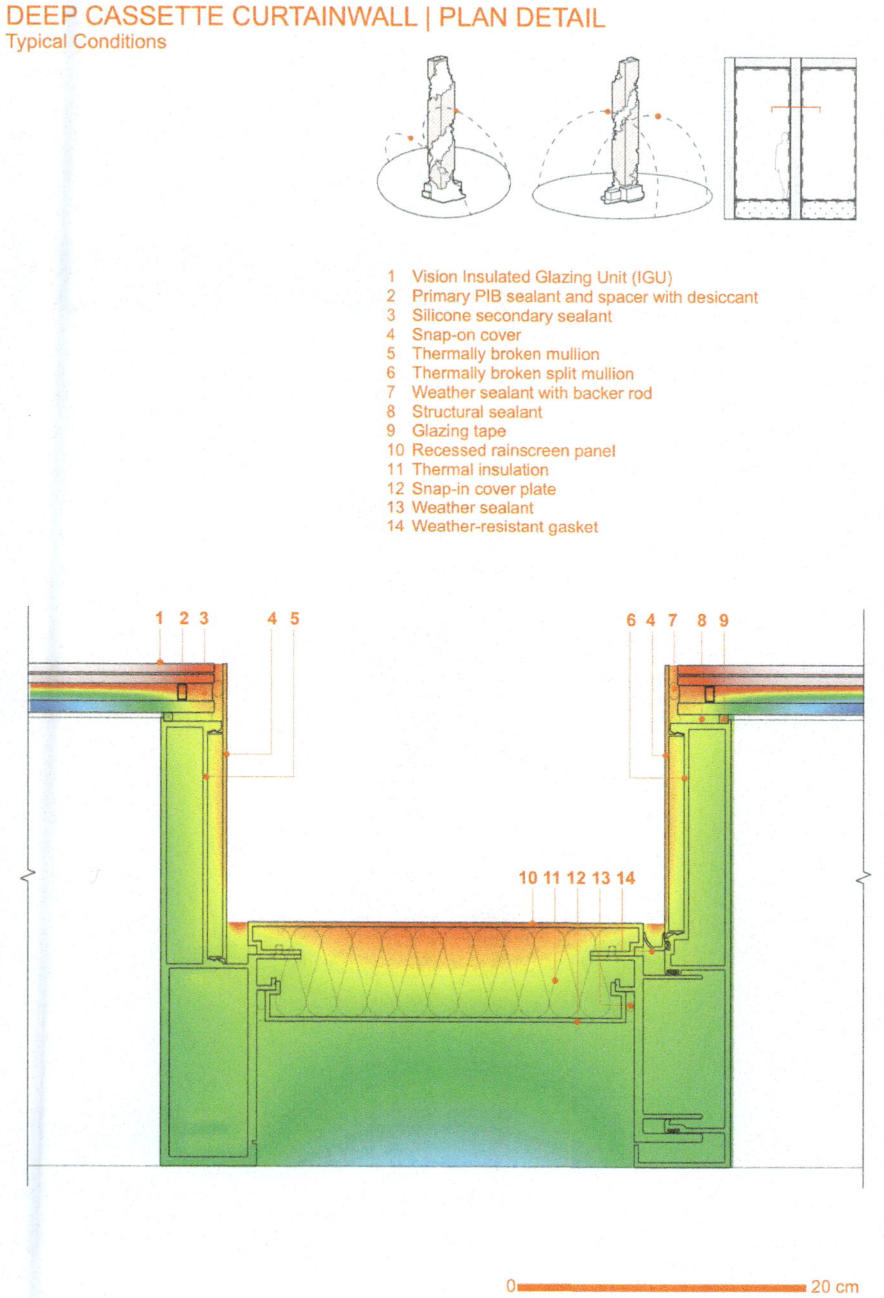

Fig. 4.37 Plan detail of a typical deep cassette curtainwall

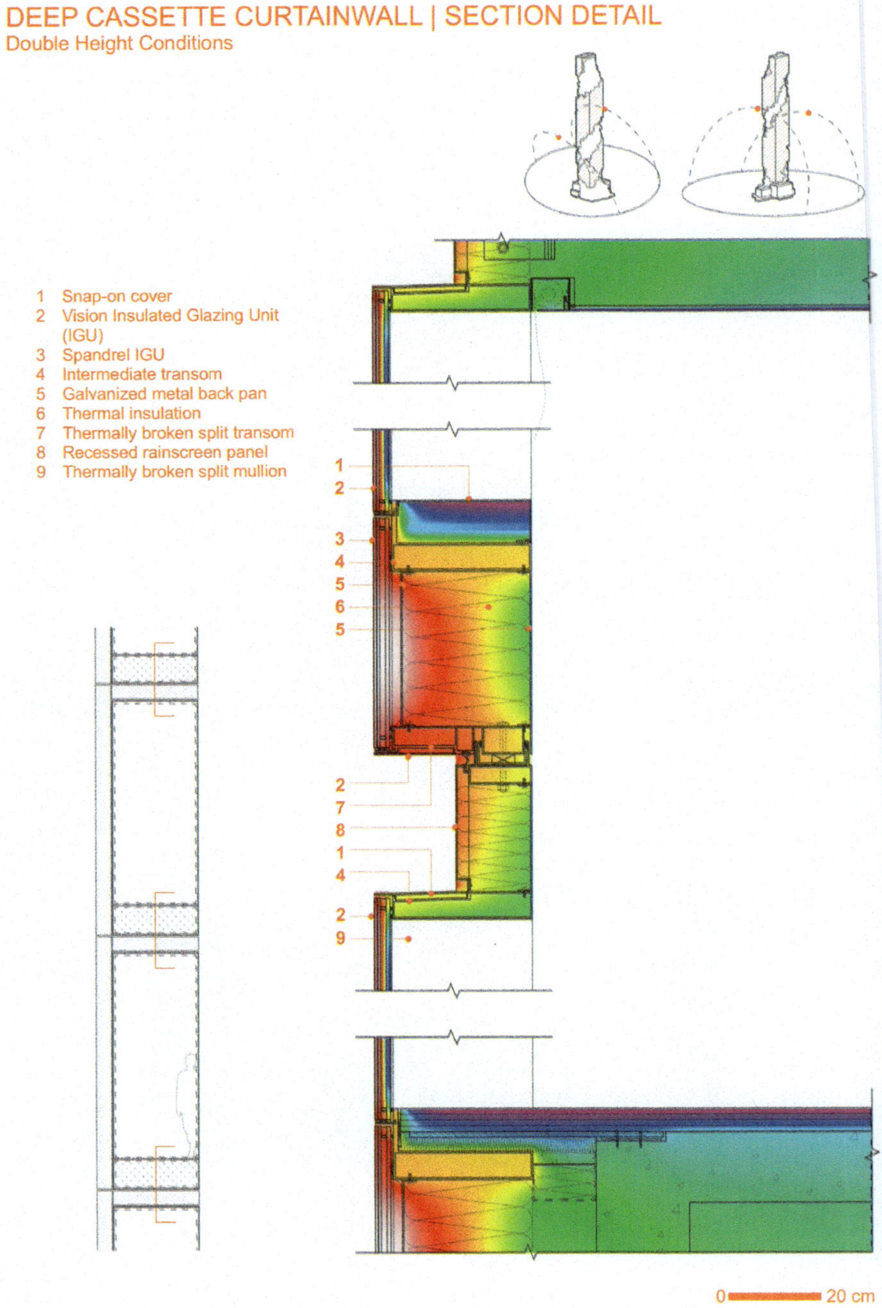

Fig. 4.38 Section detail of the deep cassette curtainwall at the intermediate transom

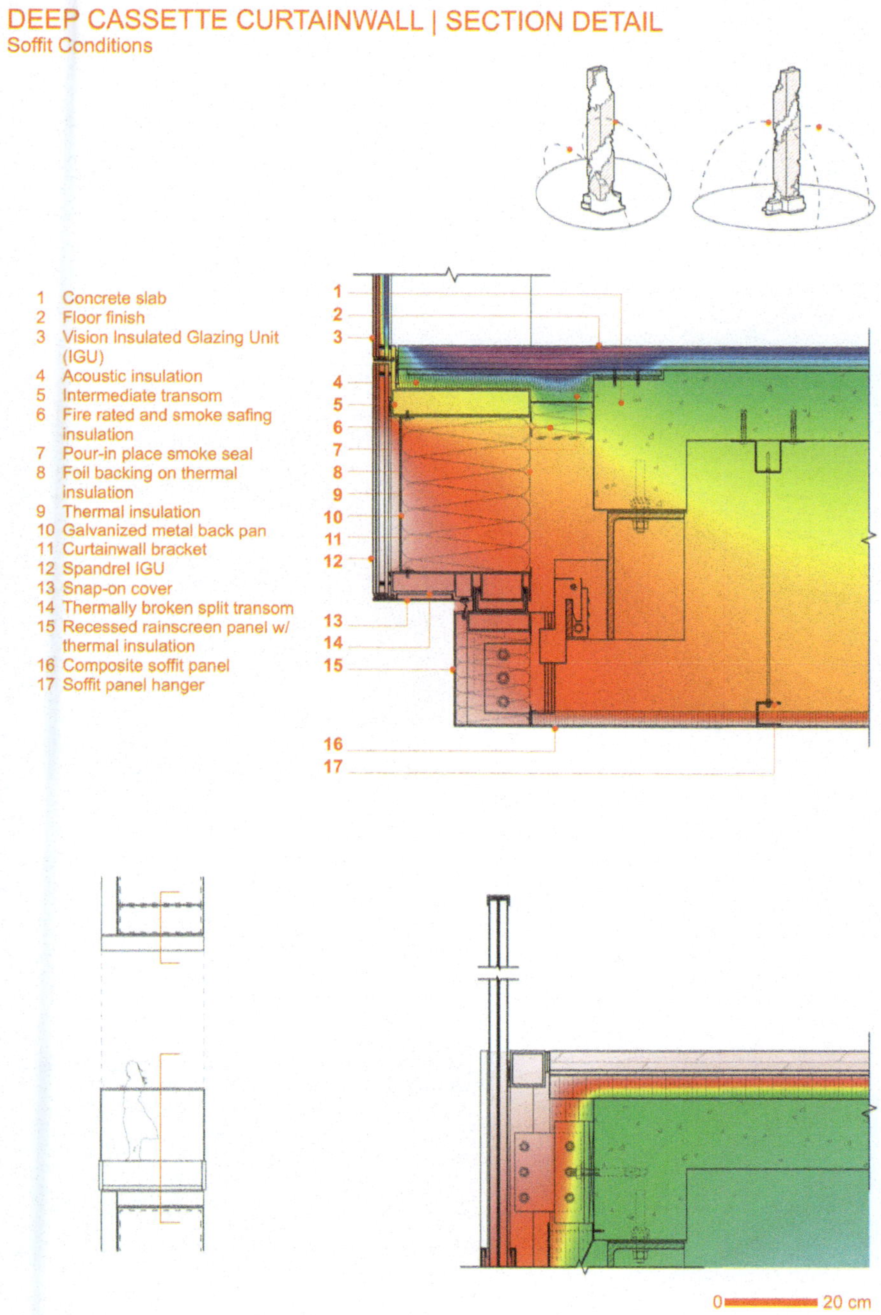

Fig. 4.39 Section detail of the deep cassette curtainwall at the soffit level

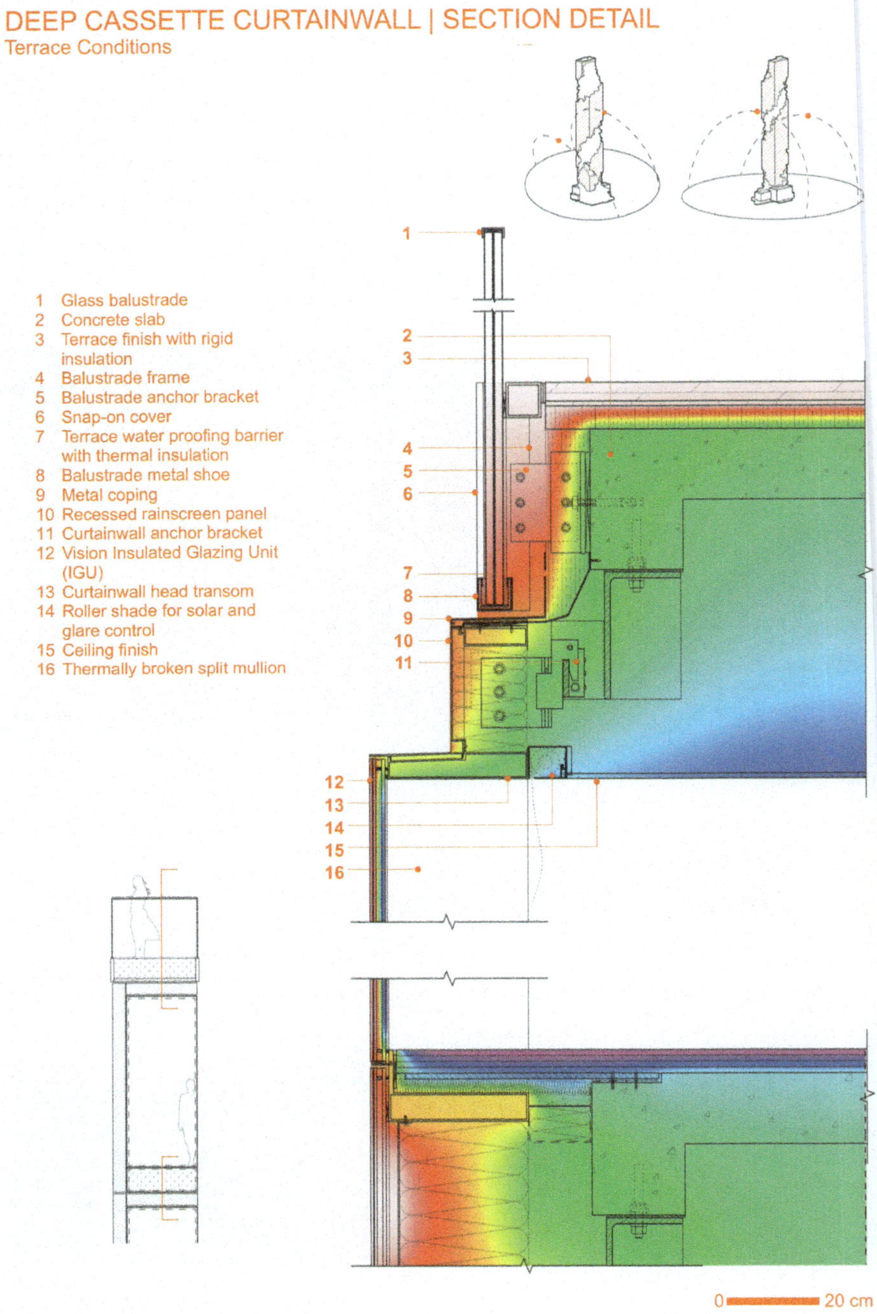

Fig. 4.40 Section detail of the deep cassette curtainwall at the terrace level

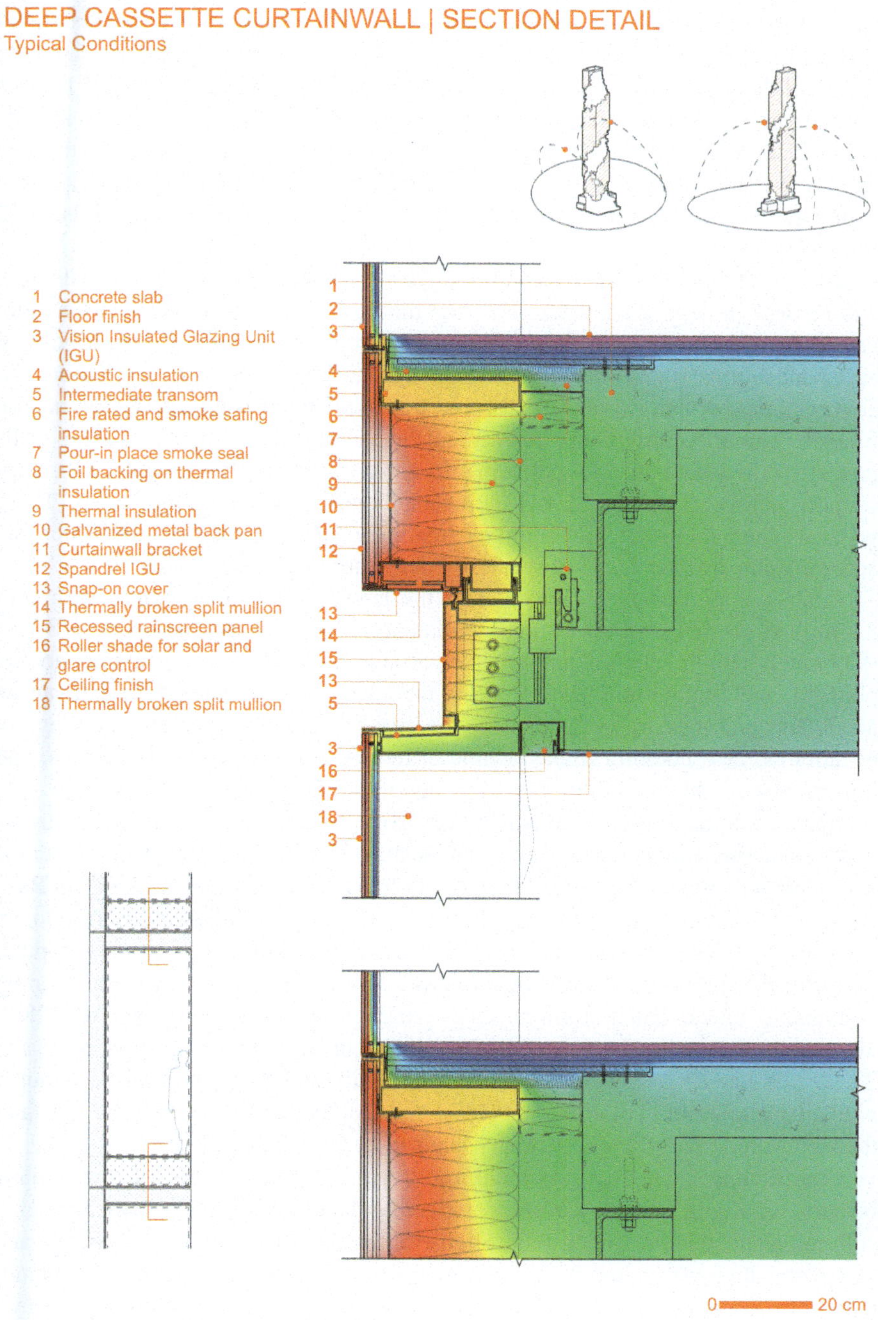

Fig. 4.41 Section detail of the deep cassette curtainwall

an approach commonly known as biophilic design principles. The towers showcase cutting-edge sustainable architecture, aiming to strengthen residents' connection with the natural environment. Tower green terraces and dense vertical façades offer tranquility within a busy cityscape. Such modern ecological techniques include green roofs and living walls, high performance technologies, renewable and low carbon systems, as well as environmentally friendly materials, which help meet contemporary sustainability standards and environmental stewardship (Figs. 4.42 and 4.43).

There is an interplay between various types of plants on green terraces, balconies and vertical gardens that collectively form this iconic ecological façade. This innovative use of common areas not only adds architectural value but also contributes to economic returns, social interaction, and environmental conservation. In high-density environments, pollutants tend to become trapped. Living walls can act as natural filters by capturing airborne pollutants and enhancing local air quality. Living walls provide vertical habitats for wildlife in urban areas, attracting insects and birds. They support pollinators through flowering plants, thereby enhancing biodiversity. These walls also help regulate temperature by creating localized microclimates, which can accommodate a wider range of species in dense urban environments.

Kuala Lumpur, Malaysia's capital, is the largest city in the country with over 8.8 million residents within its metropolitan area, covering approximately 243 km^2 (93.8 mi^2), blending natural and urban environments [287]. It is a major economic, political, and cultural hub located within the Klang Valley between the Titiwangsa Mountains and the Strait of Malacca. The city is often described as a melting pot of cultures, incorporating Malay, Chinese, Indian, and indigenous influences, and hosting a wide range of cultural festivals and traditions [288].

Kuala Lumpur's urban development revolves around enhancing its public transportation network and creating more sustainable urban spaces. Key projects include MRT (mass rapid transit) and LRT (light rail transit) networks, as well as Tun Raza Exchange financial district, among others, which aim to redevelop older, abandoned sites within the city. These initiatives are intended to support increasing population growth while positioning Kuala Lumpur on global map as one of the leading sustainable cities. Kuala Lumpur has a tropical climate (Koppen: Af) [289]. Temperatures remain relatively constant throughout year, averaging between 26 and 28 °C (79–82 °F), accompanied by high humidity and heavy rainfall year-round, with an annual rainfall of 2981 mm (117.4 in). October and November are typically the wettest months, whereas February tends to be comparatively dry.

Constructing such high-rise buildings is capital-intensive but provides many benefits for Kuala Lumpur's citizens, reflecting the city's spatial, historical, social, cultural, and economic dimensions. High-rise buildings enhance a city's image and global standing, which improves tourism opportunities while providing iconic landmarks for locals [290]. Furthermore, these buildings contribute positively to envi-

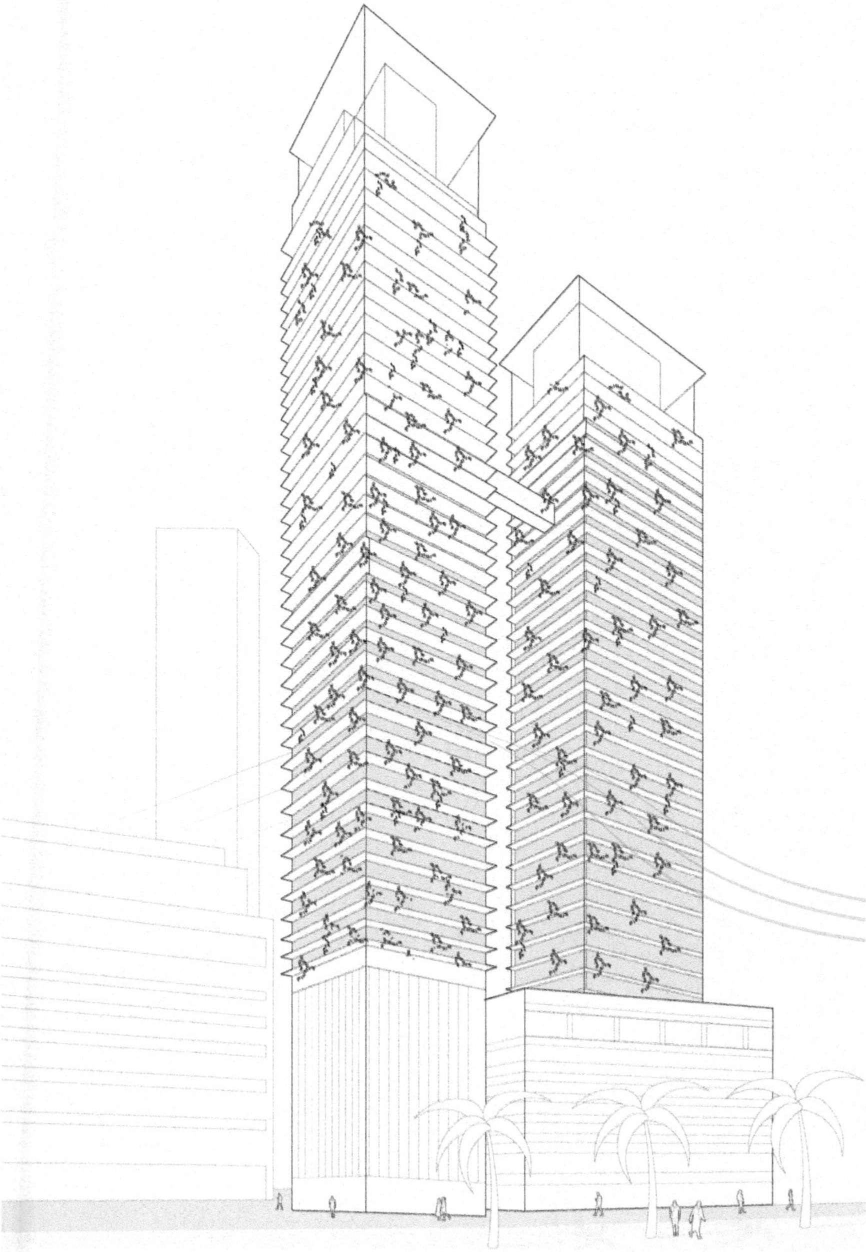

Fig. 4.42 Exterior view of Le Nouvel KLCC Kuala Lumpur, Malaysia (drawings reproduced courtesy of The Ateliers Jean Nouvel)

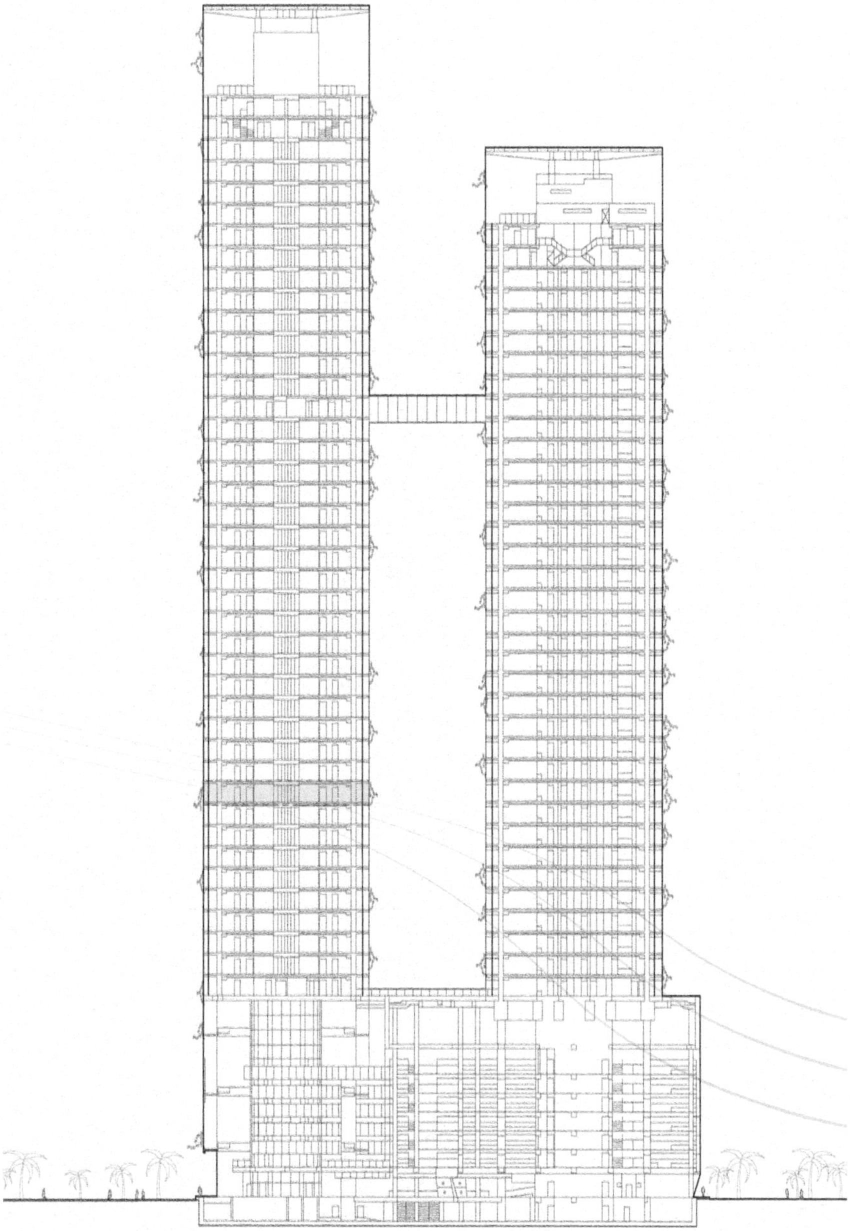

Fig. 4.43 Building section view of Le Nouvel KLCC Kuala Lumpur, Malaysia (drawings reproduced courtesy of The Ateliers Jean Nouvel)

ronmental sustainability by reducing energy consumption through the integration of green technologies and sustainable building materials. This adds value beyond what traditional low-rise buildings typically provide. Taller buildings increase floor area ratio—the total usable floor area built on a given plot of land, which can accommodate more residents compared to low-rise buildings, making efficient use of limited urban land in cities like Kuala Lumpur. This enables businesses located inside them to access larger markets while providing employment opportunities and infrastructure, leading to greater overall economic growth. Tall buildings offer economic benefits through higher rents and increased privacy, making them attractive investments in busy urban areas. These sustainable design solutions benefit both occupants and society.

Sustainable façade strategies include high-performance building enclosures incorporating automated shading devices, vegetation systems, thermal insulation, light shelves, and on-site renewable energy production. These features help reduce solar heat gain, support natural ventilation and cooling, and generate clean power, thereby enhancing indoor comfort while minimizing overall energy consumption [291]. Tall buildings require careful planning and engineering due to their scale and architectural complexity considering economic, social, cultural, and environmental implications. Local governments play a critical role in mitigating negative impacts through regulations and incentives that promote sustainable development. Because urban contexts vary significantly, each city requires specific strategies for successful high-rise development, emphasizing the importance of comprehensive research and feasibility analysis prior to project implementation [292] (Figs. 4.44, 4.45, 4.46, 4.47, 4.48, 4.49, 4.50, 4.51, 4.52, 4.53, 4.54, 4.55 and 4.56).

Fig. 4.44 Le Nouvel, Kuala Lumpur, Malaysia (2019), and sustainable façade applications

Location	Kuala Lumpur, Malaysia
Köppen Climate Classification	Tropical Rainforest Climate (Köppen: Af)
Completion Date	2016
Building Owner/Primary Tenant	Wing Tai Malaysia
Architects & Design Teams	Ateliers Jean Nouvel
Number of Floors	48 and 43 story- twin towers
Gross Floor Areas	75,000 m²

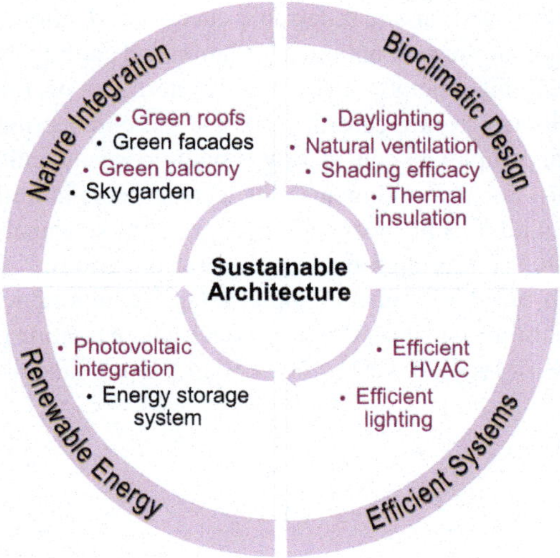

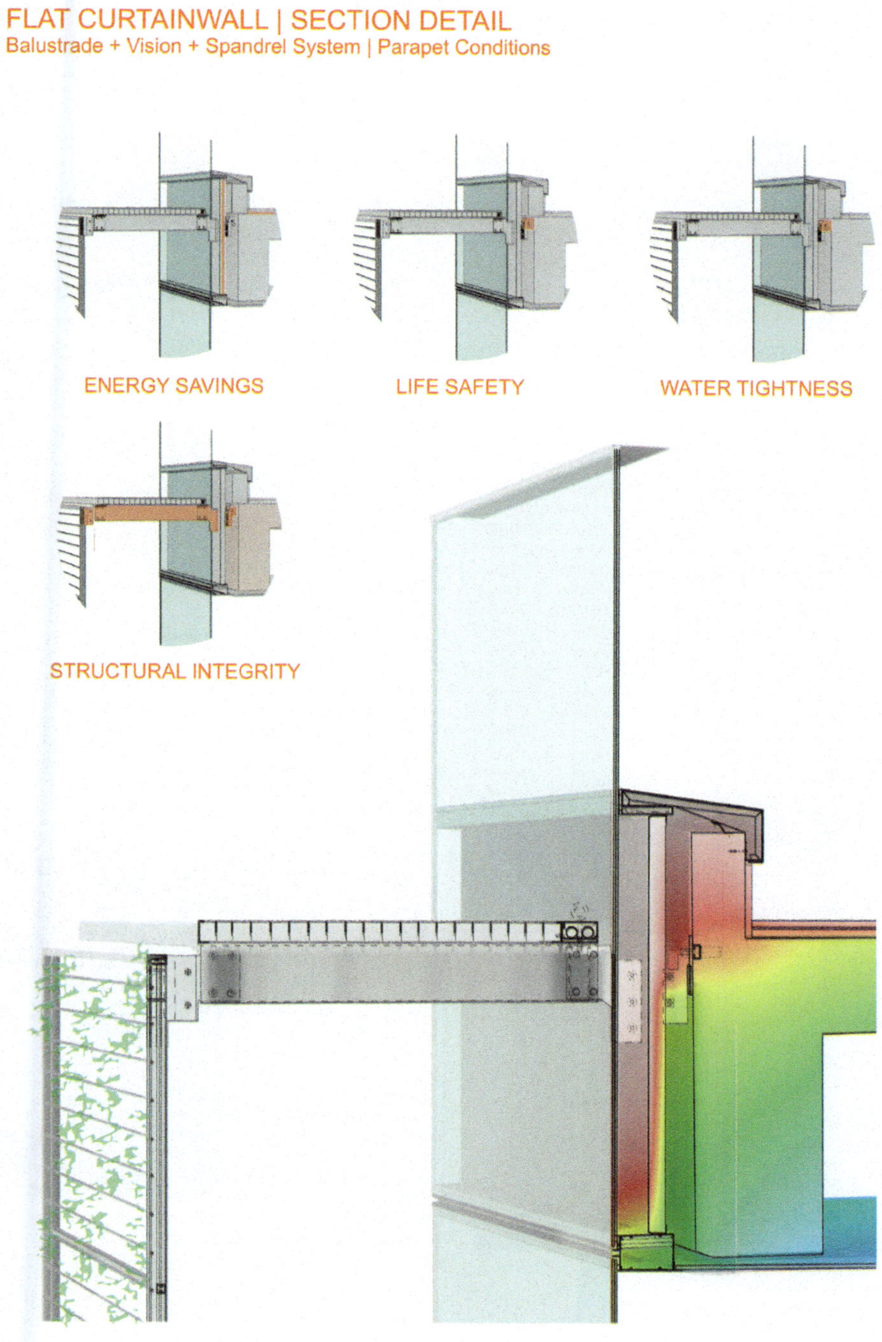

Fig. 4.45 Section details of the curtainwall at the parapet condition

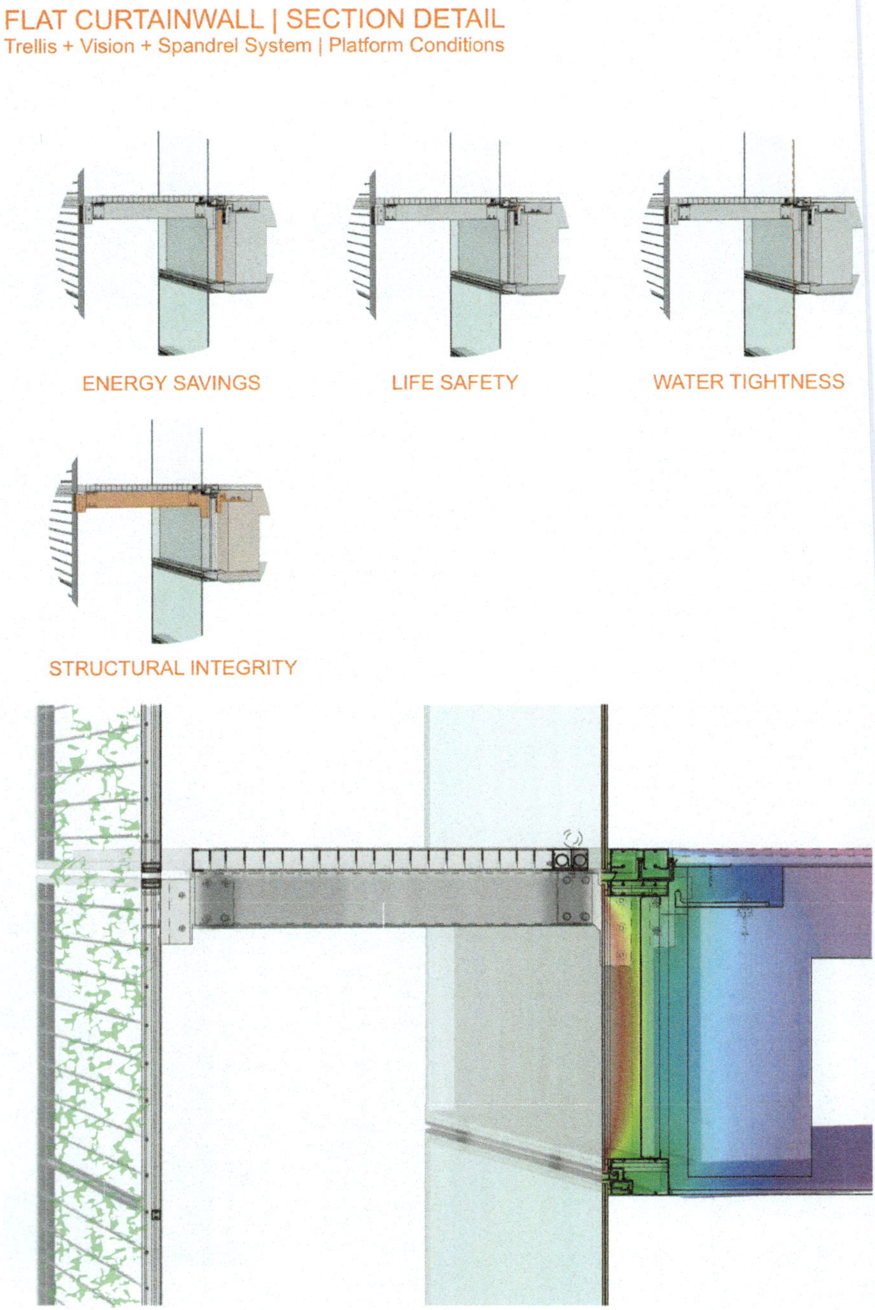

Fig. 4.46 Section details of the curtainwall with a catwalk at the spandrel

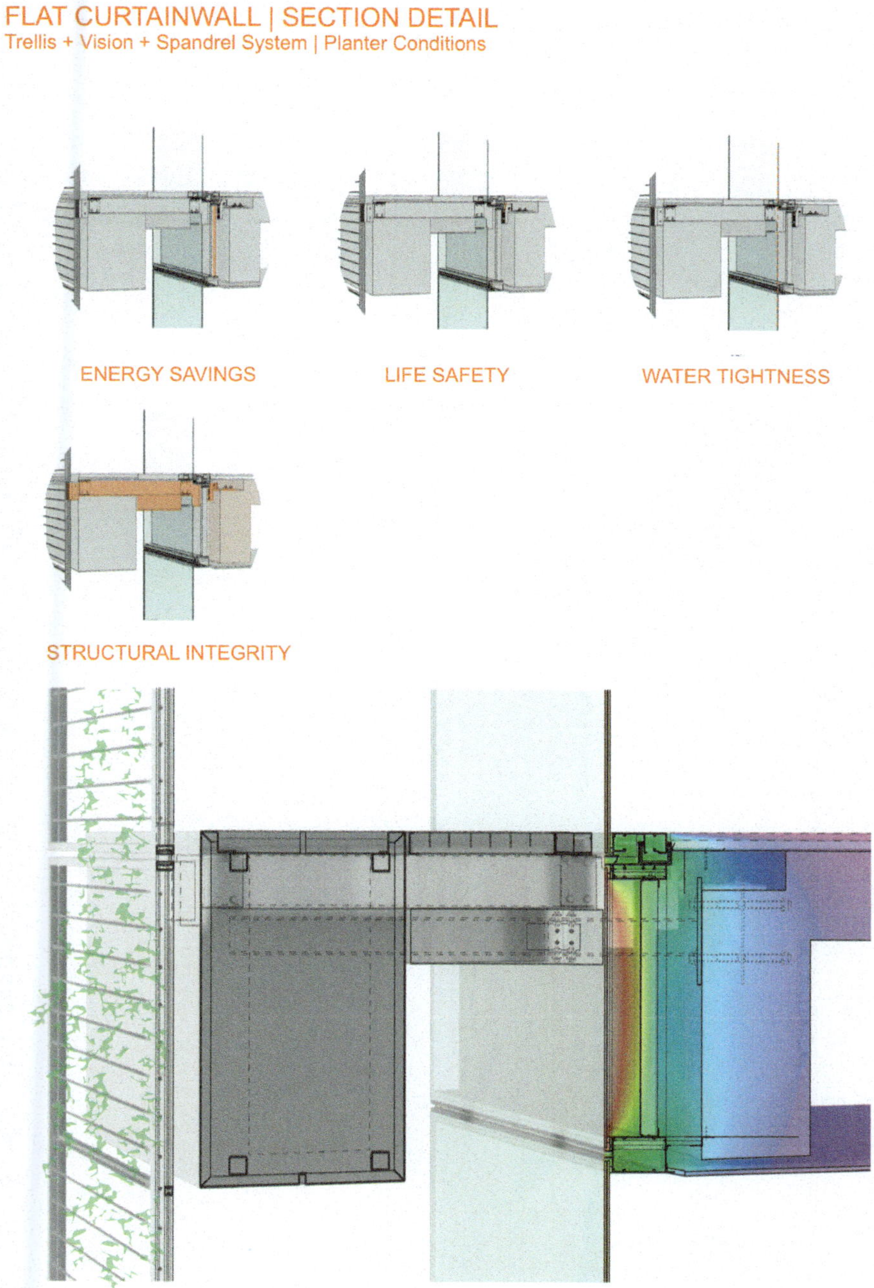

Fig. 4.47 Section details of the curtainwall with a planter at the spandrel

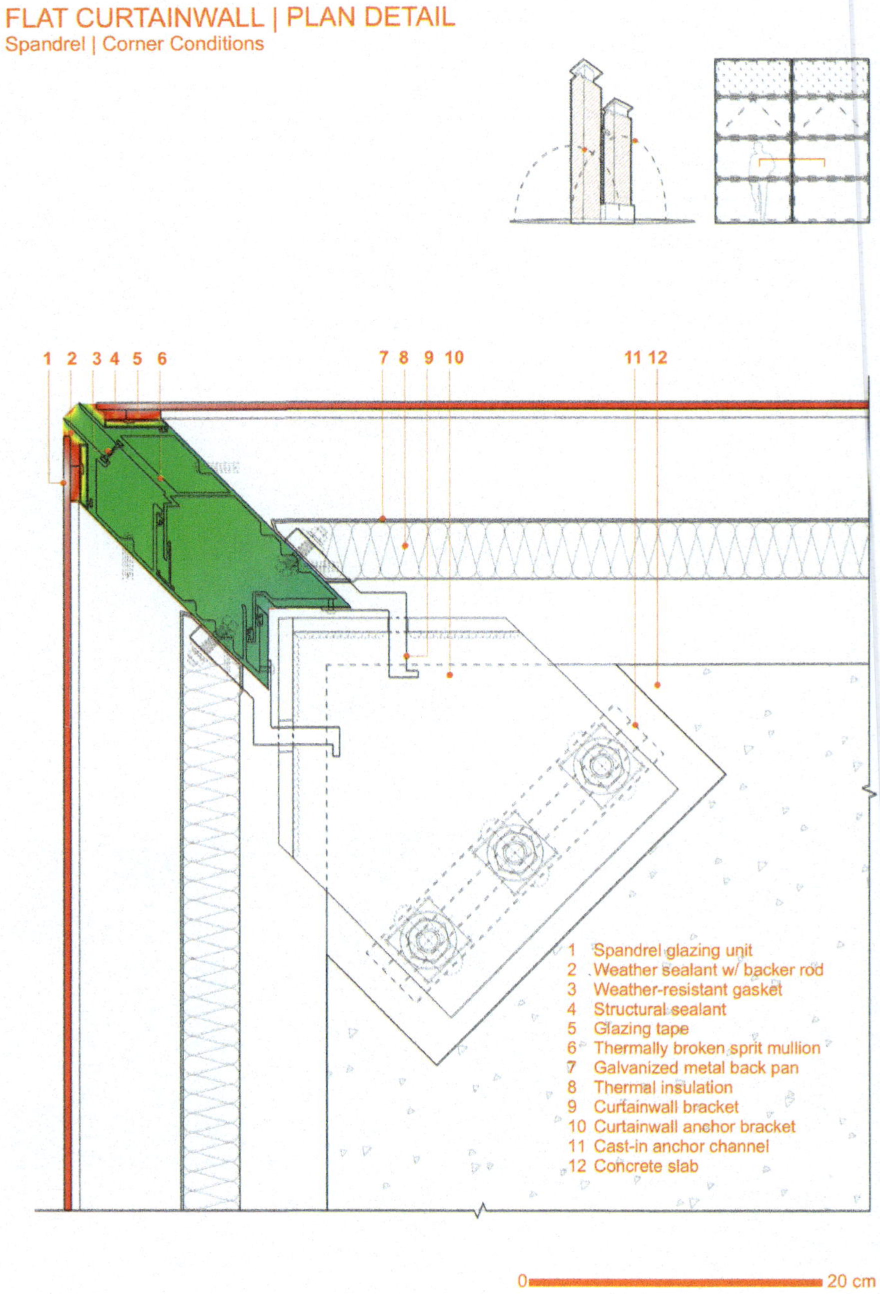

Fig. 4.48 Plan details of the curtainwall spandrel at the corner condition

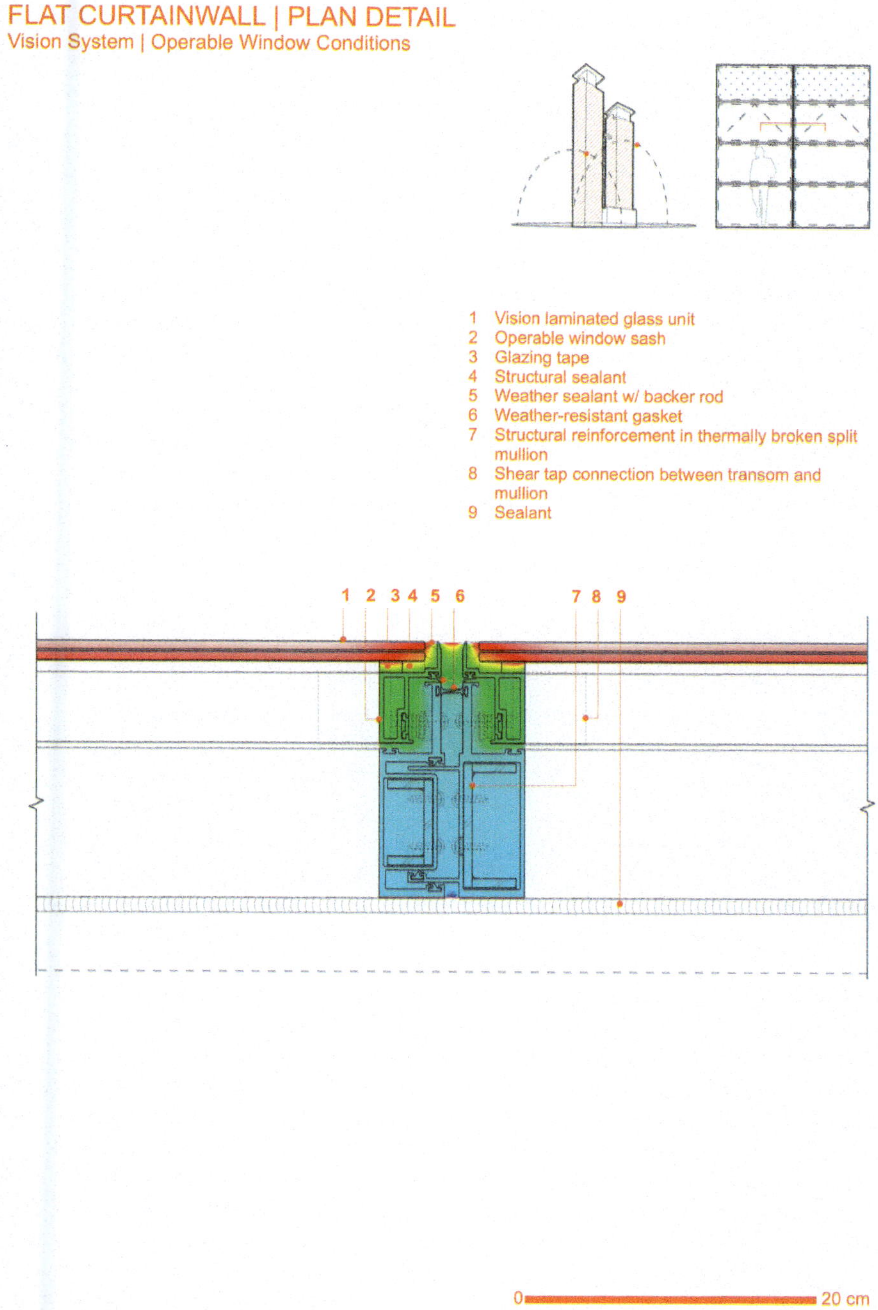

Fig. 4.49 Plan details of the curtainwall with an operable window

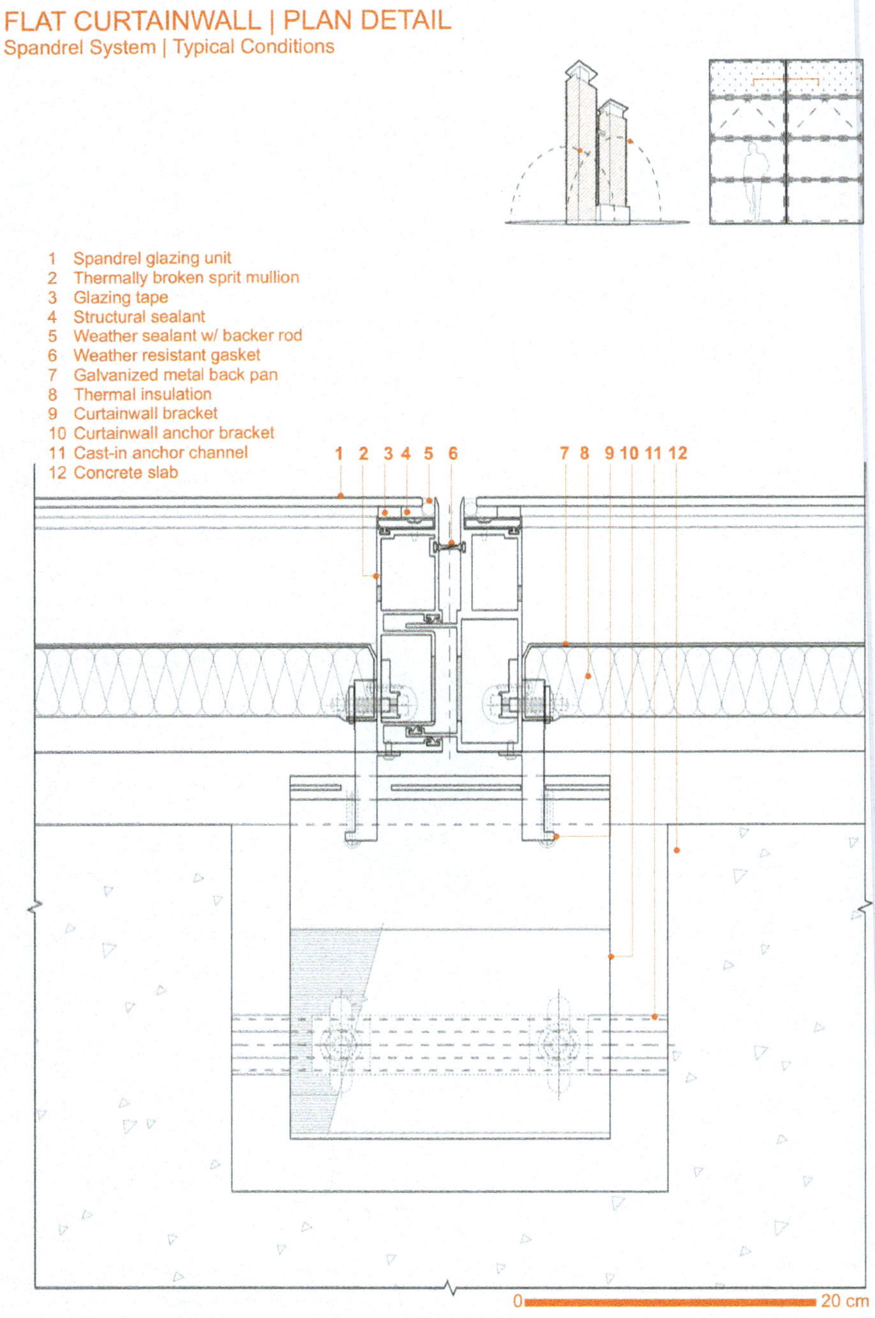

Fig. 4.50 Plan details of the curtainwall at the spandrel

FLAT CURTAINWALL | PLAN DETAIL

Vision System | Interior Wall Conditions

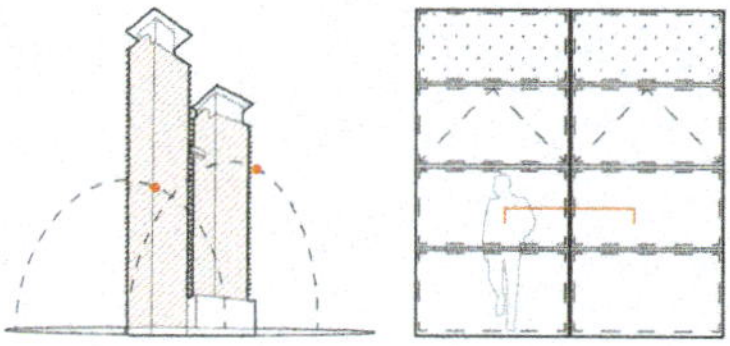

1 Vision laminated glass unit
2 Thermally broken sprit mullion
3 Glazing tape
4 Structural sealant
5 Weather sealant w/ backer rod
6 Weather-resistant gasket
7 Acoustic sealant
8 Structural VHB tape
9 Acoustic panel
10 Partition end trim
11 Fire rated and smoke safing insulation
12 Acoustic panel
13 Concrete wall
14 Partition end trim bracket

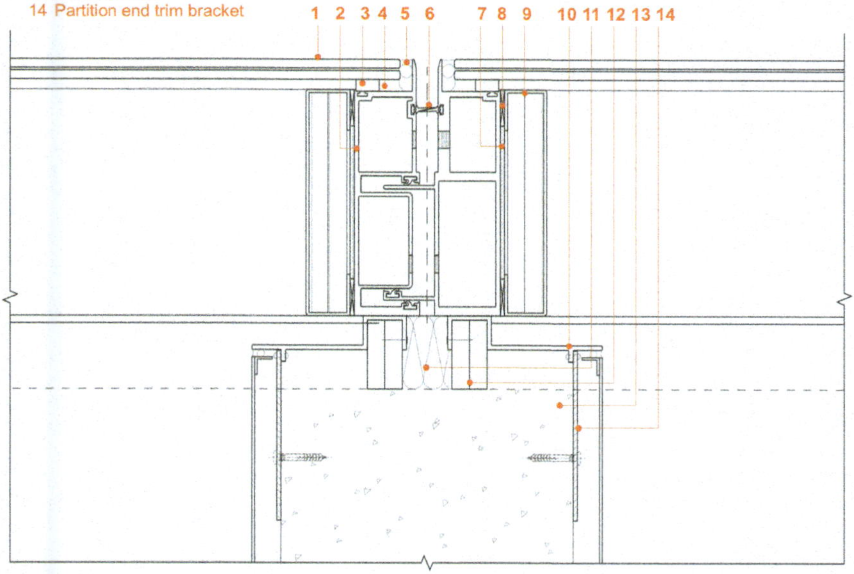

Fig. 4.51 Plan details of the curtainwall vision panel at the interior wall termination

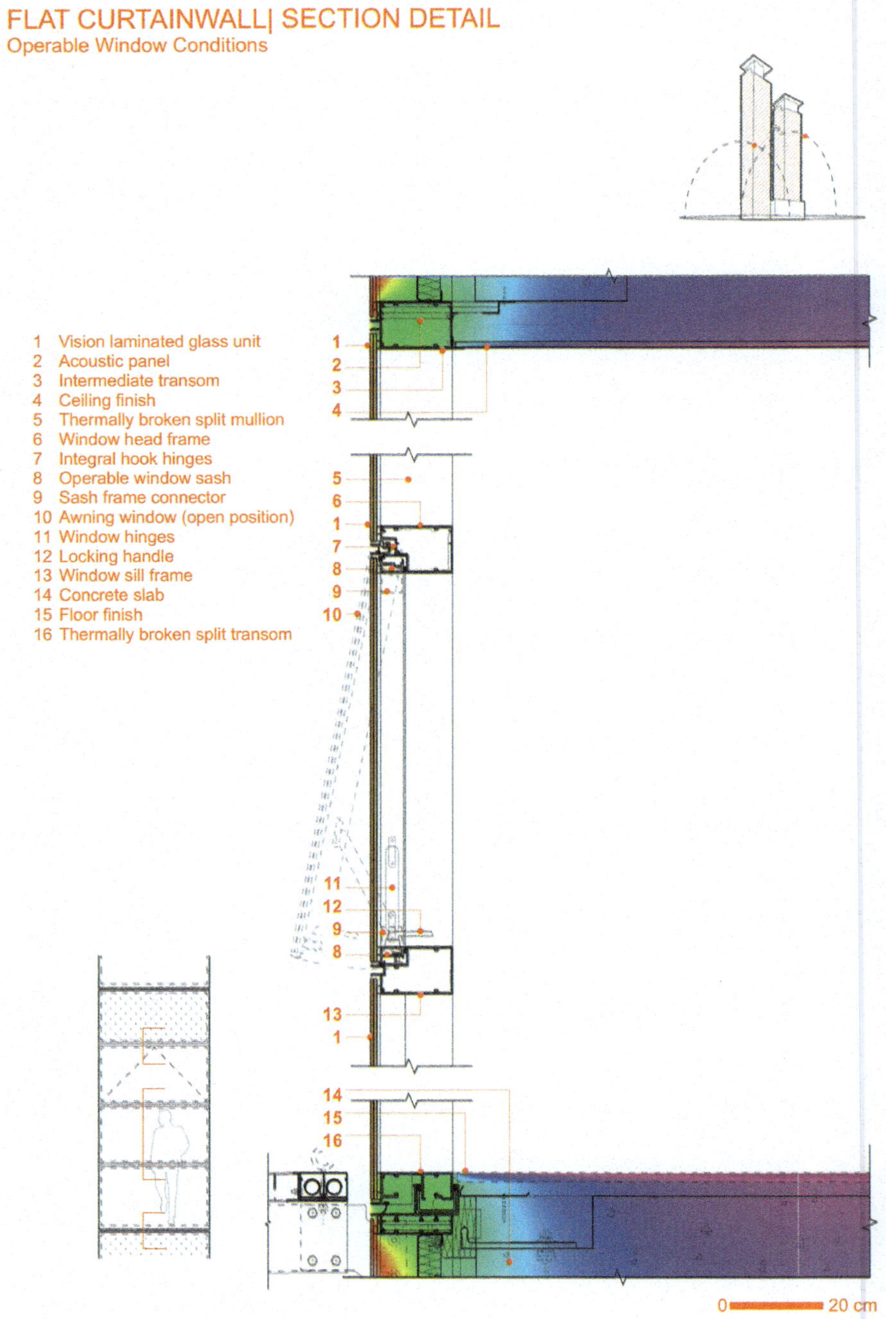

Fig. 4.52 Section details of the curtainwall with an operable window at the middle level

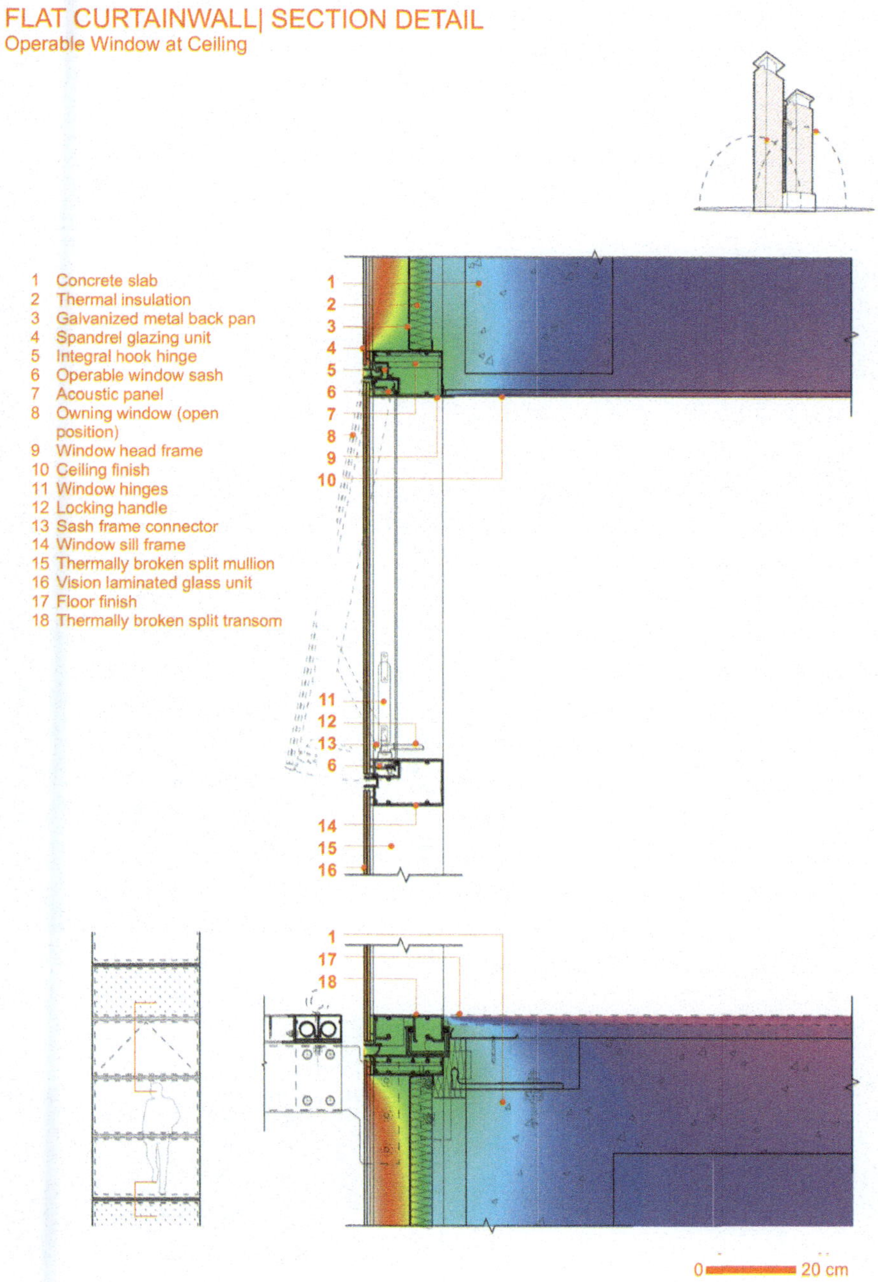

Fig. 4.53 Section details of the curtainwall with an operable window at the ceiling level

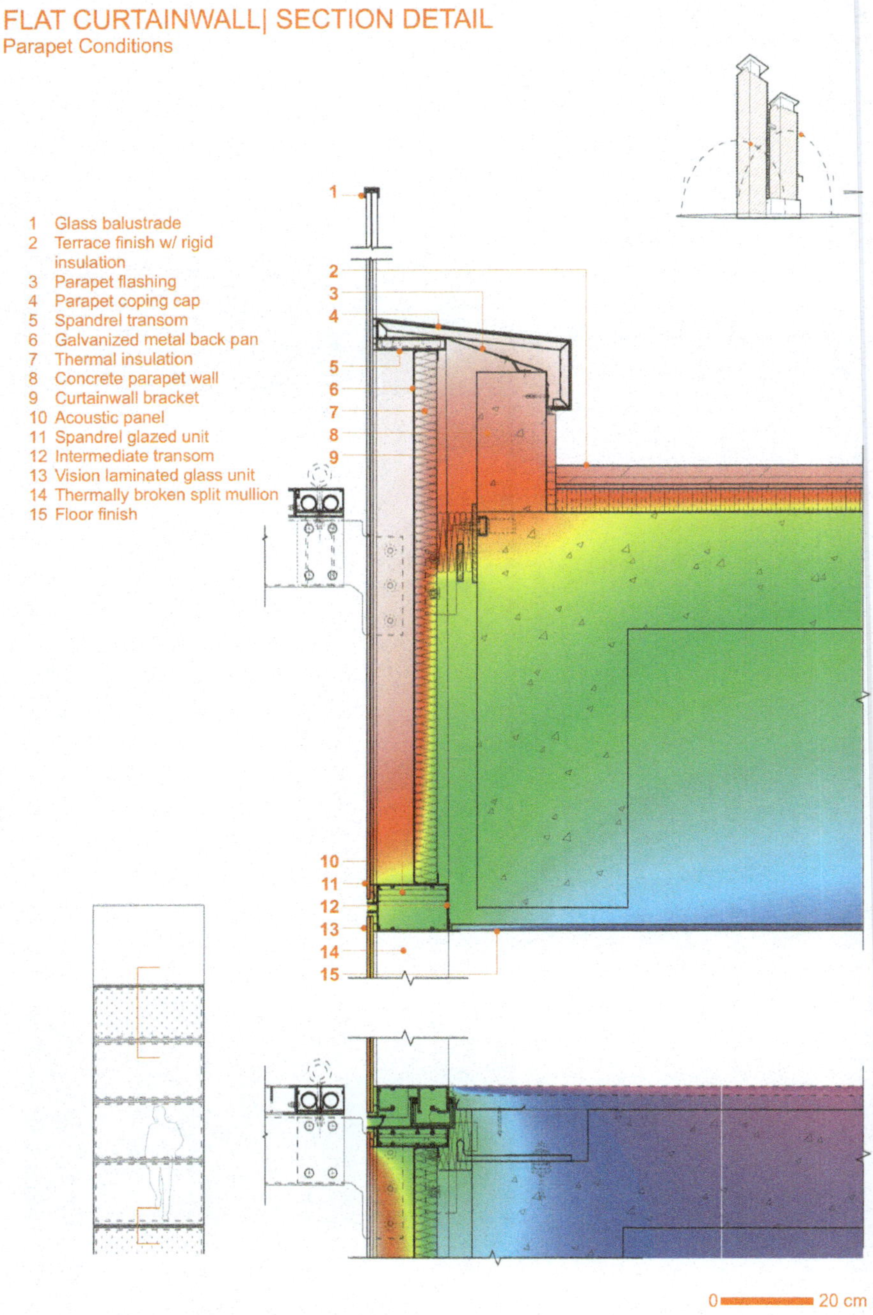

Fig. 4.54 Section details of the curtainwall at the parapet condition

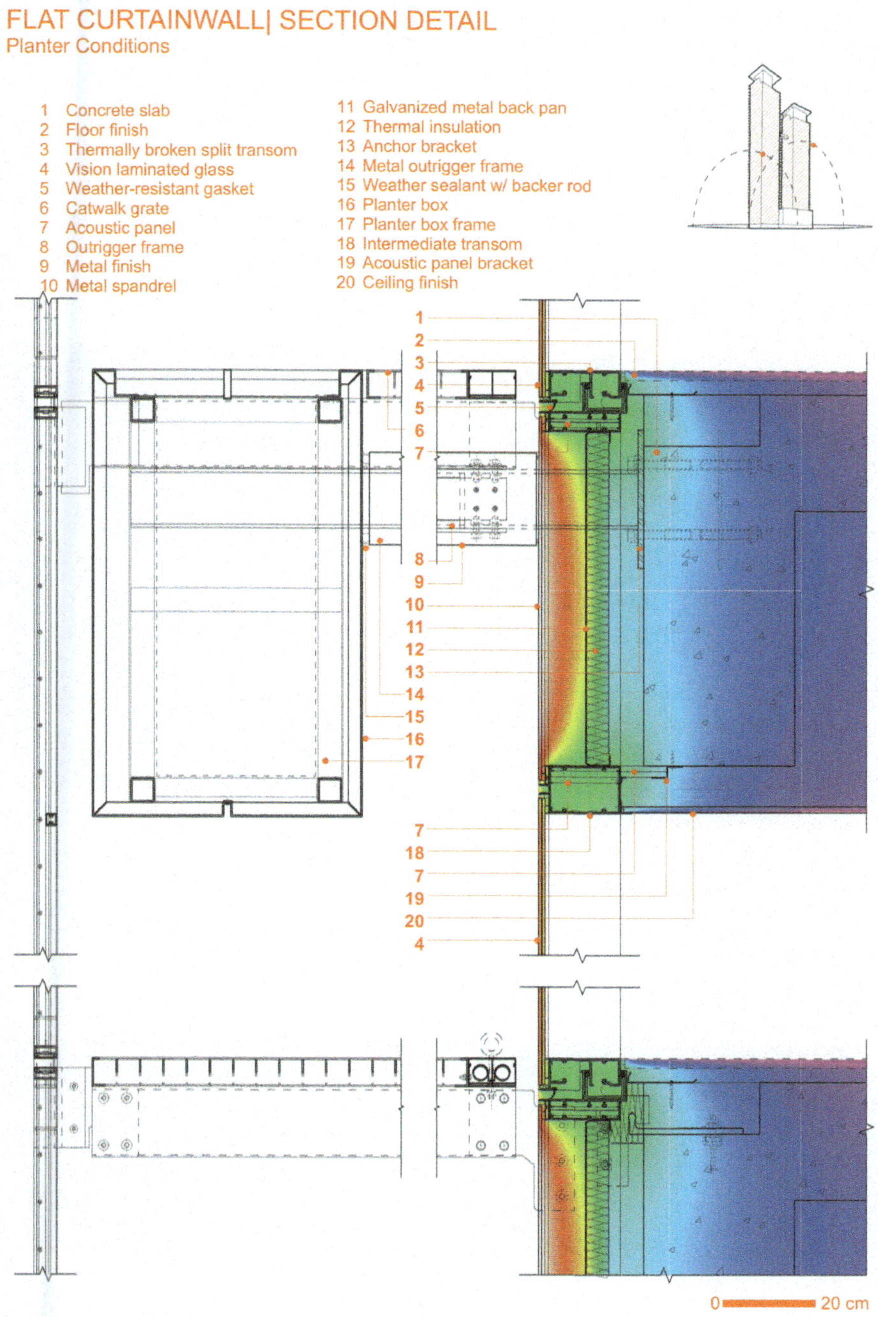

Fig. 4.55 Section details of the curtainwall with a planter and catwalk

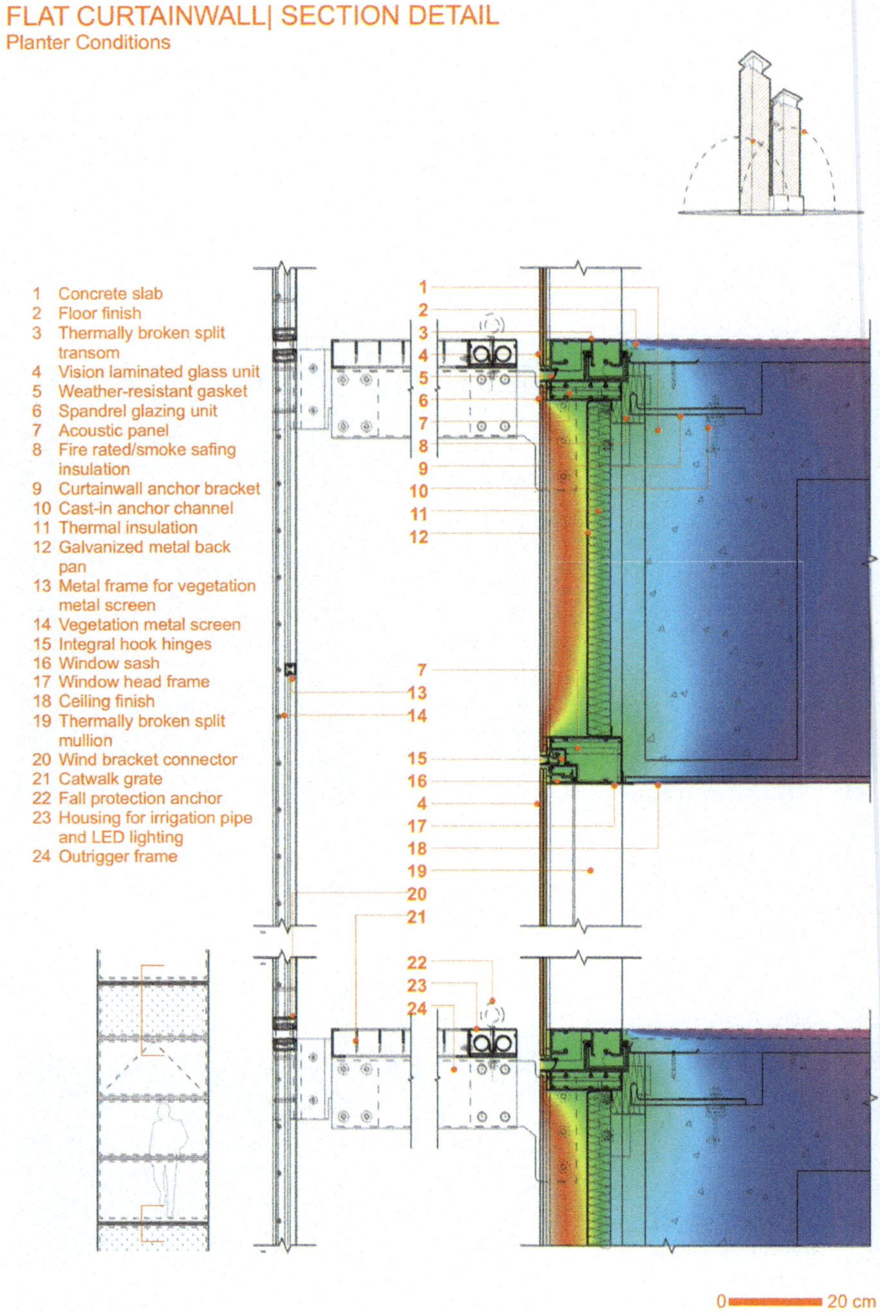

Fig. 4.56 Section details of the curtainwall with catwalk

References

1. Council on Tall Buildings and Urban Habitat (CTBUH). "Tall Building Criteria." Accessed April 27, 2024. https://www.ctbuh.org/resource/height.
2. International Code Council (ICC). *2021 International Building Code (IBC).* Accessed April 27, 2024. https://codes.iccsafe.org/content/IBC2021P2/chapter-2-definitions.
3. International Code Council (ICC). *2021 International Building Code (IBC).* Accessed April 27, 2024. https://codes.iccsafe.org/content/IBC2021P2/chapter-4-special-detailed-requirements-based-on-occupancy-and-use#IBC2021P2_Ch04_Sec403.
4. ASHRAE Technical Committee 9.12. "Tall Buildings." Accessed June 24, 2024. https://tpc.ashrae.org/?cmtKey=db0c9a13-0614-4c28-a5e6-4ad3b5e1693f.
5. International Code Council. *2018 International Building Code.* Accessed June 24, 2024. https://www.iccsafe.org/products-and-services/i-codes/2018-i-codes/ibc/.
6. Council on Tall Buildings and Urban Habitat. "Council on Tall Buildings and Urban Habitat." Accessed June 24, 2024. https://www.ctbuh.org/.
7. Emporium Hotels. "Luxury Hotels in Brisbane & South Bank | Emporium Hotels." Accessed June 24, 2024. https://www.emporiumhotels.com.au/.
8. Leicester City Council. Leicester City Council Website. Accessed June 24, 2024. https://www.leicester.gov.uk/.
9. Lexadin. "Federal Building Code." Accessed June 24, 2024. https://www.lexadin.nl/wlg/legis/nofr/eur/arch/ger/Federal-Building-Code.pdf.
10. Department of Economic and Social Affairs. "Around 2.5 Billion More People Will Be Living in Cities by 2050, Projects New UN Report." United Nations. Accessed April 27, 2024. https://www.un.org/en/desa/around-25-billion-more-people-will-be-living-cities-2050-projects-new-un-report.
11. Arora, Mohit, Felix Raspall, Lynette Cheah, and Arlindo Silva. "Buildings and the circular economy: Estimating urban mining, recovery and reuse potential of building components." *Resources, Conservation and Recycling* 154 (2020): 104581.
12. Ross, Brandon E., Cancan Yang, Michael Carlos Barrios Kleiss, Pinar Okumus, and Negar Elhami Khorasani. "Tessellated structural-architectural systems: Concept for efficient construction, repair, and disassembly." *Journal of Architectural Engineering* 26, no. 3 (2020): 04020020.
13. Ottenhaus, Lisa-Mareike, Zidi Yan, Reinhard Brandner, Paola Leardini, Gerhard Fink, and Robert Jockwer. "Design for adaptability, disassembly and reuse–A review of reversible timber connection systems." *Construction and Building Materials* 400 (2023): 132823.
14. Nawarathna, Amalka, Zaid Alwan, Barry Gledson, and Nirodha Fernando. "Embodied carbon in commercial office buildings: Lessons learned from Sri Lanka." *Journal of Building Engineering* 42 (2021): 102441.
15. Sun, Qiming, Qiong Huang, Zhuocheng Duan, and Anxiao Zhang. "Recycling potential comparison of mass timber constructions and concrete buildings: A case study in China." *Sustainability* 14, no. 10 (2022): 6174.

K. H. Kim, *Continuity in Sustainable Facades*,
https://doi.org/10.1007/978-3-032-19841-9

16. Munaro, Mayara Regina, and Sergio Fernando Tavares. "Materials passport's review: challenges and opportunities toward a circular economy building sector." *Built Environment Project and Asset Management* 11, no. 4 (2021): 767–782.
17. Karamitsos, Dimitris, Alain Schilli, and Christophe Rynikiewicz. "Servitisation for Energy Transition and Circular Economy: Learnings from the CaaS and EaaS Business Cases." In *Smart Services Summit*, pp. 53–63. Cham: Springer Nature Switzerland, 2022.
18. Kara, Hanif, Leire Asensio-Villoria, and Andreas Georgoulias, eds. *Architecture and Waste: A (re) planned Obsolescence*. Actar D, Inc., 2024. p. 46.
19. Gavu, Emmanuel Kofi, and Richard B. Peiser. "Embodied Carbon and the Nuances in Office-to-Residential Conversions." *Sustainability* 16, no. 7 (2024): 2711.
20. EIA. "Commercial Buildings ... - Energy Information Administration (EIA)." n.d. Accessed April 29, 2024. https://www.eia.gov/consumption/commercial/data/2018/.
21. Our World in Data. "Population Density by City, 2014." n.d. Accessed April 29, 2024. https://ourworldindata.org/grapher/population-density-by-city.
22. United Nations. "World Urbanization Prospects 2018 | Population Division." n.d. Accessed April 29, 2024. https://population.un.org/wup/Publications/Files/WUP2018-Highlights.pdf.
23. Skyscraper Center. "2017 Tall Building Year in Review." Accessed July 28, 2021. https://www.skyscrapercenter.com/year-in-review/2017.
24. City of New York. "Local Law 92 of 2019 and Local Law 94 of 2019: Green and Solar ..." n.d. Accessed April 29, 2024. https://www.nyc.gov/assets/buildings/pdf/green_roof_solar_ll92-n-94of2019_sn.pdf.
25. Barr, Jason M. *Building the skyline: The birth and growth of Manhattan's skyscrapers*. Oxford University Press, 2016.
26. Choi, Hee Sun, Gerhard Bruyns, Tian Cheng, Wang Zhang, and Saijal Sharma. "Human cognition and emotions for human-centred design in vertical interior urban spaces." *Journal of urban design* 30, no. 1 (2025): 114–137.
27. Gang, Jeanne. "Wanted: tall buildings less iconic, more specific." In *CTBUH 8th World Congress on Tall & Green: Typology for a Sustainable Urban Future*, pp. 496–502. 2008.
28. Zhao, Xiaoyang, Xia Mao, and Yuxiu Lu. "Skyscraper height and urban development in developing countries: economy and trade." *Kybernetes* 53, no. 4 (2024): 1505–1527.
29. Li, Songlin, Xiuyan Liu, and Qiao Wang. "Agglomeration and innovation: Evidence from skyscraper development in China." *China Economic Quarterly International* 3, no. 4 (2023): 273–283.
30. Velamati, Sri. "Feasibility, benefits and challenges of modular construction in high rise development in the United States: a developer's perspective." PhD diss., Massachusetts Institute of Technology, 2012.
31. Tse, Kam-Tim, Peter A. Hitchcock, Kenny CS Kwok, Sukit Thepmongkorn, and Chun Man Chan. "Economic perspectives of aerodynamic treatments of square tall buildings." *Journal of Wind Engineering and Industrial Aerodynamics* 97, no. 9–10 (2009): 455–467.
32. Turan, Irmak, Andrea Chegut, Daniel Fink, and Christoph Reinhart. "The value of daylight in office spaces." *Building and Environment* 168 (2020): 106503.
33. Zhu, Cindy, Andrew White, Paul Mathew, Jeff Deason, and Philip Coleman. *Raising the rent premium: Moving green building research beyond certifications and rent*. eScholarship, University of California, 2023.
34. Güneralp, Burak, Yuyu Zhou, Diana Ürge-Vorsatz, Mukesh Gupta, Sha Yu, Pralit L. Patel, Michail Fragkias, Xiaoma Li, and Karen C. Seto. "Global scenarios of urban density and its impacts on building energy use through 2050." *Proceedings of the National Academy of Sciences* 114, no. 34 (2017): 8945–8950.
35. Fragkias, Xiaoma Li, and Karen C. Seto. "Global scenarios of urban density and its impacts on building energy use through 2050." *Proceedings of the National Academy of Sciences* 114, no. 34 (2017): 8945–8950.
36. Al-Kodmany, Kheir, and Mir M. Ali. *The future of the city: Tall buildings and urban design*. WIT press, 2013.

37. Connors, John Patrick, Christopher S. Galletti, and Winston TL Chow. "Landscape configuration and urban heat island effects: assessing the relationship between landscape characteristics and land surface temperature in Phoenix, Arizona." *Landscape ecology* 28 (2013): 271–283.
38. Aram, Farshid, Ester Higueras García, Ebrahim Solgi, and Soran Mansournia. "Urban green space cooling effect in cities." *Heliyon* 5, no. 4 (2019).
39. Piracha, Awais, and Muhammad Tariq Chaudhary. "Urban air pollution, urban heat island and human health: a review of the literature." *Sustainability* 14, no. 15 (2022): 9234.
40. Piracha, Awais, and Muhammad Tariq Chaudhary. "Urban air pollution, urban heat island and human health: a review of the literature." *Sustainability* 14, no. 15 (2022): 9234.
41. Green roof and wall policy in North America. Accessed June 5, 2024. https://static1.squarespace.com/static/58e3eecf2994ca997dd56381/t/5d84dfc371cf0822bdf7dc29/1568989140101/Green_Roof_and_Wall_Policy_in_North_America.pdf.
42. A review of Green Roof Laws & Policies. Accessed June 5, 2024. https://guarinicenter.org/wp-content/uploads/2019/03/A-Review-of-Green-Roof-Laws-Policies.pdf.
43. Chicago-zoning-Ordinance.pdf. Accessed June 5, 2024. https://secondcityzoning.org/resources/Chicago-Zoning-Ordinance.pdf.
44. "Zoning Administrator (ZA) Bulletin No. 11: Better Roofs Ordinance." Zoning Administrator (ZA) Bulletin No. 11: Better Roofs Ordinance | SF Planning. Accessed June 4, 2024. https://sfplanning.org/resource/zoning-administrator-bulletin-no-11-better-roofs-ordinance.
45. Buildings news bulletin for LL92/94 of 2019. Accessed June 4, 2024. https://www.nyc.gov/assets/buildings/newsletters/DOB_BN_LL92-94_111519.html.
46. City of Toronto. "City of Toronto Green Roof Bylaw." City of Toronto, July 4, 2022. https://www.toronto.ca/city-government/planning-development/official-plan-guidelines/green-roofs/green-roof-bylaw/.
47. Beam plus new buildings v2.0 (2023 edition) Manual. Accessed June 4, 2024. https://www.beamsociety.org.hk/en/BEAM-Plus/-/media/7D5637F9105048C98896A99790925B6D.ashx.
48. "Tax Incentives, Funding Assistance & Green Mortgage." HKGBC. Accessed June 4, 2024. https://www.hkgbc.org.hk/eng/beam-plus/incentive-funding/index.jsp.
49. "About." Skyrise Greenery. Accessed June 4, 2024. https://skyrisegreenery.nparks.gov.sg/incentive-scheme/about/.
50. Seoulsolution. "Roof Gardening Support Project for Private Buildings." 서울정책아카이브 Seoul Solution, May 2, 2017. https://seoulsolution.kr/en/content/roof-gardening-support-project-private-buildings.
51. My covenant. Accessed June 5, 2024. http://mycovenant.eumayors.eu/storage/web/mc_covenant/documents/9/rb8kPdegeeeJC7R3n0D1ZzLru4Tcd9nX.xlsx.
52. Why do we need the urban forest fund? - city of Melbourne. Accessed June 4, 2024. https://www.melbourne.vic.gov.au/community/greening-the-city/urban-forest-fund/Pages/why-need-urban-forest-fund.aspx.
53. Lauermann, John. "Vertical gentrification: A 3D analysis of luxury housing development in New York City." *Annals of the American Association of Geographers* 112, no. 3 (2022): 772–780.
54. Jones, Craig E., and David Ley. "Transit-oriented development and gentrification along Metro Vancouver's low-income SkyTrain corridor." *The Canadian Geographer/Le Géographe Canadien* 60, no. 1 (2016): 9–22.
55. Jacobs, Jane. "The death and life of great American cities. 1961." *New York: Vintage* 321, no. 325 (1992): 9783839413272-099.
56. Newman, Oscar. *Creating defensible space*. Diane Publishing, 1997.
57. Barros, Paula, Linda Ng Fat, Leandro MT Garcia, Anne Dorothee Slovic, Nikolas Thomopoulos, Thiago Herick de Sa, Pedro Morais, and Jennifer S. Mindell. "Social consequences and mental health outcomes of living in high-rise residential buildings and the influence of planning, urban design and architectural decisions: A systematic review." *Cities* 93 (2019): 263–272.

58. Barros, Paula, Linda Ng Fat, Leandro MT Garcia, Anne Dorothee Slovic, Nikolas Thomopoulos, Thiago Herick de Sa, Pedro Morais, and Jennifer S. Mindell. "Social consequences and mental health outcomes of living in high-rise residential buildings and the influence of planning, urban design and architectural decisions: A systematic review." *Cities* 93 (2019): 263–272.
59. Barros, Paula, Linda Ng Fat, Leandro MT Garcia, Anne Dorothee Slovic, Nikolas Thomopoulos, Thiago Herick de Sa, Pedro Morais, and Jennifer S. Mindell. "Social consequences and mental health outcomes of living in high-rise residential buildings and the influence of planning, urban design and architectural decisions: A systematic review." *Cities* 93 (2019): 263–272.
60. Graham, Stephen, and Lucy Hewitt. "Getting off the ground: On the politics of urban verticality." *Progress in Human Geography* 37, no. 1 (2013): 72–92.
61. Bahadure, Sarika, and Rajashree Kotharkar. "Social sustainability and mixed land use." In *Proceedings of International Conference on Advances in Architecture and Civil Engineering (AARCV 2012)*, vol. 21, p. 725. 2012.
62. Viñoly, Rafael, C. Blomberg, and M. Blanes. "Challenges and benefits of integrating public space into tall buildings." *CTBUH Research Paper* (2015): 282–287.
63. Ibraeva, Anna, Gonçalo Homem de Almeida Correia, Cecília Silva, and António Pais Antunes. "Transit-oriented development: A review of research achievements and challenges." *Transportation Research Part A: Policy and Practice* 132 (2020): 110–130.
64. Ibraeva, Anna, Gonçalo Homem de Almeida Correia, Cecília Silva, and António Pais Antunes. "Transit-oriented development: A review of research achievements and challenges." *Transportation Research Part A: Policy and Practice* 132 (2020): 110–130.
65. Admin. "Safety Features in World's Tallest Building - Burj Khalifa." Apaha Institute of Construction Project Management, June 24, 2019. https://apah.org.in/safety-features-in-worlds-tallest-building-burj-khalifa/.
66. Shanghai Tower. Accessed June 6, 2024. https://aiahouston.org/media/uploads/resource-docs/ao11.pdf.
67. Izudinshah Abd. Wahab, Lokman Hakim Ismail, and Aslila Abd Kadir. "Opening Design and Position Effect on Building Natural Stack Effect and Cross Ventilation." *International Journal of Engineering and Technical Research* 5, no. 1 (May 2016).
68. Izudinshah Abd. Wahab, Lokman Hakim Ismail, and Aslila Abd Kadir. "Opening Design and Position Effect on Building Natural Stack Effect and Cross Ventilation." *International Journal of Engineering and Technical Research* 5, no. 1 (May 2016).
69. Zitao Jiang, Tomohiro Kobayashi, Toshio Yamanaka, and Mats Sandberg, "A Literature Review of Cross Ventilation in Buildings," *Energy and Buildings* 291 (2023): 113143, accessed June 14, 2024, https://doi.org/10.1016/j.enbuild.2023.113143.
70. Zitao Jiang, Tomohiro Kobayashi, Toshio Yamanaka, and Mats Sandberg, "A Literature Review of Cross Ventilation in Buildings," *Energy and Buildings* 291 (2023): 113143, accessed June 14, 2024, https://doi.org/10.1016/j.enbuild.2023.113143.
71. Ghaffarianhoseini Ali, Amirhosein Ghaffarianhoseini, Umberto Berardi, John Tookey, Danny Hin Wa Li, and Shahab Kariminia. "Exploring the Advantages and Challenges of Double-Skin Façades (DSFs)." *Renewable and Sustainable Energy Reviews* 60 (2016): 1052–1065. https://doi.org/10.1016/j.rser.2016.01.130.
72. Ghaffarianhoseini Ali, Amirhosein Ghaffarianhoseini, Umberto Berardi, John Tookey, Danny Hin Wa Li, and Shahab Kariminia. "Exploring the Advantages and Challenges of Double-Skin Façades (DSFs)." *Renewable and Sustainable Energy Reviews* 60 (2016): 1052–1065. https://doi.org/10.1016/j.rser.2016.01.130.
73. Veitch, J. A., and R. Gifford. 1996. "Choice, Perceived Control, and Performance Decrements in the Physical Environment." Journal of Environmental Psychology 16 (3): 269–276.
74. Becker, F., and F. Steele. 1995. *Workplace by Design: Mapping the High-Performance Workscape.* San Francisco, CA: Jossey-Bass.
75. Mardaljevic, John, Lisa Heschong, and Eleanor Lee. "Daylight Metrics and Energy Savings." *Lighting Research + Technology* 41, no. 3 (2009): 261–283.

76. Mardaljevic, John, Lisa Heschong, and Eleanor Lee. "Daylight Metrics and Energy Savings." *Lighting Research + Technology* 41, no. 3 (2009): 261–283.
77. Garkuwa, Jamilu, Adel Abdou, and Muhammad Asif. "Dynamic Facades for Sustainable Buildings: A Review of Classification, Applications, Prospects and Challenges." *Energy Reports* 11 (2024): 5999–6014. https://doi.org/10.1016/j.egyr.2024.05.047.
78. Ozdemir, Aydin. "The effect of window views' openness and naturalness on the perception of rooms' spaciousness and brightness: A visual preference study." *Scientific Research and Essays* 5, no. 16 (2010): 2275–2287.
79. Pastore, Luisa, and Marilyne Andersen. "The Influence of Façade and Space Design on Building Occupants' Indoor Experience." *Journal of Building Engineering* 46 (2022): 103663. https://doi.org/10.1016/j.jobe.2021.103663.
80. Leadership in Energy and Environmental Design, "LEED," accessed June 18, 2024, https://www.usgbc.org/leed.
81. Balaras, C.A. "The Role of Thermal Mass on the Cooling Load of Buildings: An Overview of Computational Methods." *Energy and Buildings* 24, no. 1 (1996): 1–10. https://doi.org/10.1016/0378-7788(95)00956-6.
82. Kuczyński, T., and A. Staszczuk. "Experimental Study of the Influence of Thermal Mass on Thermal Comfort and Cooling Energy Demand in Residential Buildings." *Energy* 195 (2020): 116984. ISSN 0360-5442. https://doi.org/10.1016/j.energy.2020.116984.
83. Santamouris, M. "Cooling the Cities – A Review of Reflective and Green Roof Mitigation Technologies to Fight Heat Island and Improve Comfort in Urban Environments." Solar Energy 103 (2014): 682–703. https://doi.org/10.1016/j.solener.2012.07.003.
84. Galagoda, R.U., Jayasinghe, G.Y., Halwatura, R.U., & Rupasinghe, H.T. "The Impact of Urban Green Infrastructure as a Sustainable Approach towards Tropical Micro-climatic Changes and Human Thermal Comfort." *Urban Forestry & Urban Greening* 34 (2018): 1–9. ISSN 1618-8667. https://doi.org/10.1016/j.ufug.2018.05.008.
85. Bahar, Heymi. "Renewables 2017." *IEA Report, Tokyo* (2017).
86. Bahar, Heymi. "Renewables 2017." *IEA Report, Tokyo* (2017).
87. Binz, Christian, Tian Tang, and Joern Huenteler. "Spatial lifecycles of cleantech industries–The global development history of solar photovoltaics." *Energy Policy* 101 (2017): 386–402.
88. Benemann, Joachim, Oussama Chehab, and Eric Schaar-Gabriel. "Building-integrated PV modules." *Solar Energy Materials and Solar Cells* 67, no. 1–4 (2001): 345–354.
89. Liu, Zhe, Sarah E. Sofia, Hannu S. Laine, Michael Woodhouse, Sarah Wieghold, Ian Marius Peters, and Tonio Buonassisi. "Revisiting thin silicon for photovoltaics: a technoeconomic perspective." *Energy & Environmental Science* 13, no. 1 (2020): 12–23.
90. Gul, Mehreen, Yash Kotak, and Tariq Muneer. "Review on recent trend of solar photovoltaic technology." *Energy Exploration & Exploitation* 34, no. 4 (2016): 485–526.
91. Gul, Mehreen, Yash Kotak, and Tariq Muneer. "Review on recent trend of solar photovoltaic technology." *Energy Exploration & Exploitation* 34, no. 4 (2016): 485–526.
92. Gul, Mehreen, Yash Kotak, and Tariq Muneer. "Review on recent trend of solar photovoltaic technology." *Energy Exploration & Exploitation* 34, no. 4 (2016): 485–526.
93. Sopian, and Nowshad Amin. "An overview of solar photovoltaic panels' end-of-life material recycling." Energy Strategy Reviews 27 (2020): 100431.
94. Peter, L. M. "Towards Sustainable Photovoltaics: The Search for New Materials." Philosophical Transactions: Mathematical, Physical and Engineering Sciences 369, no. 1942 (2011): 1840–56. http://www.jstor.org/stable/41148899.
95. Moore, Andrew. "The Solar Lifecycle Test." ReNew: Technology for a Sustainable Future, no. 109 (2009): 46–49. https://www.jstor.org/stable/renetechsustfutu.109.46.
96. Sopian, and Nowshad Amin. "An overview of solar photovoltaic panels' end-of-life material recycling." Energy Strategy Reviews 27 (2020): 100431.
97. Italos, Christos & Patsias, Michalis & Yiangou, Andriani & Stavrinou, Stylianos & Vassiliades, Constantinos. (2022). Use of double skin façade with building integrated solar systems for an energy renovation of an existing building in Limassol, Cyprus: Energy performance analysis. Energy Reports. 8. 15144–15161. https://doi.org/10.1016/j.egyr.2022.11.088.

98. Devetaković, Mirjana, Djordje Djordjević, Milan Radojević, Aleksandra Krstić-Furundžić, Bogdan-Gabriel Burduhos, Georgios Martinopoulos, Mircea Neagoe, and Gabriele Lobaccaro. "Photovoltaics on landmark buildings with distinctive geometries." *Applied Sciences* 10, no. 19 (2020): 6696.
99. Betancur, Juan. "Multitasking façade: how to combine BIPV with passive solar mitigation strategies in a high-rise curtain wall system." *International Journal of High-Rise Buildings* 6, no. 4 (2017): 307–313.
100. Mangone, Giancarlo, S. R. Kurvers, and P. G. Luscuere. "Constructing thermal comfort: investigating the effect of vegetation on indoor thermal comfort through a four season thermal comfort quasi-experiment." *Building and environment* 81 (2014): 410–426.
101. Ulrich, Roger S. "View through a window may influence recovery from surgery." *Science* 224, no. 4647 (1984): 420–421.
102. Elsadek, Mohamed, Binyi Liu, and Junfang Xie. "Window view and relaxation: Viewing green space from a high-rise estate improves urban dwellers' wellbeing." *Urban Forestry & Urban Greening* 55 (2020): 126846.
103. Ulrich, Roger S. "View through a window may influence recovery from surgery." *science* 224, no. 4647 (1984): 420–421.
104. Yeo, Lee Bak. "Psychological And Physiological Benefits Of Plants In The Indoor Environment: A Mini And In-Depth Review." *International Journal of Built Environment and Sustainability* 8, no. 1 (2021): 57–67.
105. Dzhambov, Angel M., Peter Lercher, Matthew HEM Browning, Drozdstoy Stoyanov, Nadezhda Petrova, Stoyan Novakov, and Donka D. Dimitrova. "Does greenery experienced indoors and outdoors provide an escape and support mental health during the COVID-19 quarantine?." *Environmental Research* (2020): 110420.
106. Pérez-Urrestarazu, Luis, Maria P. Kaltsidi, Panayiotis A. Nektarios, Georgios Markakis, Vivian Loges, Katia Perini, and Rafael Fernández-Cañero. "Particularities of having plants at home during the confinement due to the COVID-19 pandemic." *Urban Forestry & Urban Greening* (2020): 126919.
107. van den Bogerd, Nicole, S. Coosje Dijkstra, Sander L. Koole, Jacob C. Seidell, and Jolanda Maas. "Greening the room: A quasi-experimental study on the presence of potted plants in study rooms on mood, cognitive performance, and perceived environmental quality among university students." *Journal of Environmental Psychology* 73 (2021): 101557.
108. Elsadek, Mohamed, Binyi Liu, and Junfang Xie. "Window view and relaxation: Viewing green space from a high-rise estate improves urban dwellers' wellbeing." *Urban Forestry & Urban Greening* 55 (2020): 126846.
109. Radić, Mina, Marta Brković Dodig, and Thomas Auer. "Green Facades and Living Walls—A Review Establishing the Classification of Construction Types and Mapping the Benefits." *Sustainability*, published August 23, 2019. https://doi.org/10.3390/su11174579.
110. Manouchehri, M., Santiago López, J., and Valiente López, M. "Sustainable Design of Vertical Greenery Systems: A Comprehensive Framework." *Sustainability* 16 (2024): 3249. https://doi.org/10.3390/su16083249.
111. Velazquez-Gonzalez, R.S., Garcia-Garcia, A.L., Ventura-Zapata, E., Barceinas-Sanchez, J.D.O., and Sosa-Savedra, J.C. "A Review on Hydroponics and the Technologies Associated for Medium- and Small-Scale Operations." *Agriculture* 12, no. 5 (2022): 646. https://doi.org/10.3390/agriculture12050646.
112. Kour, K., Gupta, D., Gupta, K., Dhiman, G., Juneja, S., Viriyasitavat, W., Mohafez, H., and Islam, M.A. "Smart-Hydroponic-Based Framework for Saffron Cultivation: A Precision Smart Agriculture Perspective." *Sustainability* 14 (2022): 1120. https://doi.org/10.3390/su14031120.
113. Pomoni, D.I., Koukou, M.K., Vrachopoulos, M.G., and Vasiliadis, L. "A Review of Hydroponics and Conventional Agriculture Based on Energy and Water Consumption, Environmental Impact, and Land Use." Energies 16, no. 4 (2023): 1690. https://doi.org/10.3390/en16041690.

114. Sambo, P., Nicoletto, C., Giro, A., Pii, Y., Valentinuzzi, F., Mimmo, T., Lugli, P., Orzes, G., Mazzetto, F., Astolfi, S., Terzano, R., and Cesco, S. "Hydroponic Solutions for Soilless Production Systems: Issues and Opportunities in a Smart Agriculture Perspective." Frontiers in Plant Science 10 (2019): 923. https://doi.org/10.3389/fpls.2019.00923.
115. Reference-Medl, A., Stangl, R., and Florineth, F. "Vertical Greening Systems: A Review on Recent Technologies and Research Advancement." *Building and Environment* 125 (2017): 227–239.
116. Manso, M., and Castro-Gomes, J. "Green Wall Systems: A Review of Their Characteristics." *Renewable and Sustainable Energy Reviews* 41 (2015): 863–871.
117. Wagner, Jennifer, and Sean Monkman. "14. Making an Impact: Sustainable Success Stories: CarbonCure." *How to Commercialize Chemical Technologies for a Sustainable Future* (2021): 255–262.
118. Das, Swagato, Purnachandra Saha, Swatee Prajna Jena, and Pratyush Panda. "Geopolymer concrete: Sustainable green concrete for reduced greenhouse gas emission–A review." *Materials Today: Proceedings* 60 (2022): 62–71.
119. Heller, Patrick, Partha Mukhopadhyay, and Michael Walton. "Cabal City: Regime Theory and Indian Urbanization." *Watson Institute for International and Public Affairs Research Paper* 2016–32 (2016).
120. Heller, Patrick, Partha Mukhopadhyay, and Michael Walton. "Cabal City: Regime Theory and Indian Urbanization." *Watson Institute for International and Public Affairs Research Paper* 2016–32 (2016).
121. *Climate-data.org*. Mumbai climate: Average Temperature by month, Mumbai water temperature. (n.d.). https://en.climate-data.org/asia/india/maharashtra/mumbai-29/.
122. Pragyan Dash, Shanta, and Deepika Shetty. "Cultural Identity in Sustainable Architecture." *International Research Journal on Advanced Science Hub* 2, no. 7 (2020): 155–158.
123. Jakarta population 2024. Accessed July 6, 2024. https://worldpopulationreview.com/world-cities/jakarta-population.
124. Suswandari, Suswandari. "Local History of Jakarta and MulticulturalAttitude (Historical Local Study of Betawi Ethnic)." *Journal of Education, Teaching and Learning* 2, no. 1 (2017): 34–41.
125. *Climate-data.org*. Mumbai climate: Average Temperature by month, Mumbai water temperature. (n.d.). https://en.climate-data.org/asia/india/maharashtra/mumbai-29/.
126. Sukada, Nabila Qirala, and Purnama Salura. "Paul Rudolph's design principles on high-rise office buildings in Indonesia case study: Wisma Dharmala Sakti Jakarta and Wisma Dharmala Sakti Surabaya." *Riset Arsitektur (RISA)* 2, no. 04 (2018): 376–393.
127. Effendi, M. F., I. F. Ridzqo, and S. W. Dharmatanna. "An Overview of High-Rise Buildings in Jakarta since 1967 to 2020." In *IOP Conference Series: Earth and Environmental Science*, vol. 933, no. 1, p. 012001. IOP Publishing, 2021.
128. Effendi, M. F., I. F. Ridzqo, and S. W. Dharmatanna. "An Overview of High-Rise Buildings in Jakarta since 1967 to 2020." In *IOP Conference Series: Earth and Environmental Science*, vol. 933, no. 1, p. 012001. IOP Publishing, 2021.
129. *BGC-Ortigas Link Bridge and its benefits*. Crown Asia. (2024, March 4). https://www.crownasia.com.ph/lifestyle-blog/bgc-ortigas-link-bridge-what-does-it-bring-to-taguig-city/.
130. *Climate-data.org*. Manila climate: Average Temperature by month, Manila water temperature. (n.d.). https://en.climate-data.org/asia/philippines/manila/manila-1952/.
131. Daiichi Properties. "One World Place - Premium Grade A Building." Accessed July 18, 2023. https://daiichiproperties.com/one-world-place/.
132. Beirut population 2024. Accessed July 6, 2024. https://worldpopulationreview.com/world-cities/beirut-population.
133. Traboulsi, Fawwaz. A history of modern Lebanon. Pluto Press, 2012.
134. Traboulsi, Fawwaz. A history of modern Lebanon. Pluto Press, 2012.
135. *Climate-data.org*. Search results. (n.d.). https://en.climate-data.org/search/?q=Beirut.
136. "347 Beirut Terraces." Herzog & de Meuron. Accessed July 5, 2024. https://www.herzogdemeuron.com/projects/347-beirut-terraces/.

137. Golasz-Szolomicka, Hanna, and Jerzy Szolomicki. "Vertical Gardens in High-Rise Buildings–Modern Form of Green Building Technology." In *IOP Conference Series: Materials Science and Engineering*, vol. 603, no. 2, p. 022067. IOP Publishing, 2019.
138. *Home: Hong kong tourism board*. Discover Hong Kong. (n.d.). https://www.discoverhongkong.com/eng/index.html.
139. Population of cities in Hong Kong 2024. Accessed July 6, 2024. https://worldpopulationreview.com/countries/cities/hong-kong.
140. *Climate-data.org*. Search results. (n.d.-a). https://en.climate-data.org/search/?q=hong%2Bkong.
141. "Cheung Fai Building, Hong Kong, 2013–2015 : [MVRDV]." A + U: Architecture and Urbanism, no. 4 (April 1, 2017): 108–111. https://search.ebscohost.com/login.aspx?direct=true&AuthType=shib&db=bvh&AN=767165&authtype=shib&site=ehost-live&scope=site.
142. Singapore population 2024. Accessed July 6, 2024. https://worldpopulationreview.com/world-cities/singapore-population.
143. Cahyadi, Gundy, Barbara Kursten, Marc Weiss, and Guang Yang. "Singapore's economic transformation." *Global Urban Development: Singapore Metropolitan Economic Strategy Report* (2004): 2–25.
144. *Climate-data.org*. Singapore Climate: Weather Singapore & Temperature By Month. (n.d.). https://en.climate-data.org/asia/singapore-107/.
145. Abdel, Hana. "18 Robinson/KPF." ArchDaily, January 29, 2021. https://www.archdaily.com/955916/18-robinson-building-kpf.
146. "Guidelines for Landscape Replacement Areas within New Developments in (Part) Downtown Core, (Part) Straits View, (Part) Kallang and (Part) Jurong East Planning Areas." Urban Redevelopment Authority. Accessed July 5, 2024. https://www.ura.gov.sg/Corporate/Guidelines/Circulars/dc09-09.
147. Nguyen, Thanh Bao, D. Ary A. Samsura, Erwin van der Krabben, and Anh-Duc Le. "Saigon-ho chi minh city." *Cities* 50 (2016): 16–27.
148. *Climate-data.org*. Climate Ho Chi Minh City: Temperature, climate graph, Climate table for Ho Chi Minh City. (n.d.). https://en.climate-data.org/asia/vietnam/ho-chi-minh-city-840/.
149. VTN3. "Nanoco Head Office." VTN3. Accessed July 18, 2023. https://vtnarchitects.net/nanoco-head-office-pe187.html.
150. Ahmedabad population 2024. Accessed July 6, 2024. https://worldpopulationreview.com/world-cities/ahmedabad-population.
151. Mahadevia, Darshini, Renu Desai, and Suchita Vyas. "City Profile: Ahmedabad." *CUE working paper; 26* (2014).
152. *Climate-data.org*. Ahmedabad climate: Weather Ahmedabad & temperature by month. (n.d.). https://en.climate-data.org/asia/india/gujarat/ahmedabad-2828/.
153. "Montecarlo HQ, Ahmedabad, Gujarat - Project by Edifice." EDIFICE. Accessed July 5, 2024. https://www.edifice.co.in/Project/Montecarlo-HQ.
154. "Montecarlo HQ, Ahmedabad, Gujarat - Project by Edifice." EDIFICE. Accessed July 5, 2024. https://www.edifice.co.in/Project/Montecarlo-HQ.
155. "Santiago, Chile Metro Area Population (1950–2025)." MacroTrends. Accessed December 21, 2025. https://www.macrotrends.net/global-metrics/cities/20439/santiago/population.
156. Schuster-Olbrich, Juan Pablo, Oriol Marquet, Carme Miralles-Guasch, and Luis Fuentes Arce. "Spatial patterns and drivers of urban expansion: An exploratory spatial analysis of the Metropolitan Region of Santiago, Chile, from 1997 to 2013." Cities 153 (2024): 105305.
157. Schuster-Olbrich, Juan Pablo, Oriol Marquet, Carme Miralles-Guasch, and Luis Fuentes Arce. "Spatial patterns and drivers of urban expansion: An exploratory spatial analysis of the Metropolitan Region of Santiago, Chile, from 1997 to 2013." Cities 153 (2024): 105305.
158. *Climate-data.org*. Search results. (n.d.-a). https://en.climate-data.org/south-america/chile/region-metropolitana-de-santiago/santiago-938/.
159. Fracalossi, I. (2009, January 21). *Edificio Consorcio Santiago/Enrique Browne + Borja Huidobro*. ArchDaily en Español. https://www.archdaily.cl/cl/02-14392/edificio-consorcio-sede-santiago-enrique-browne-borja-huidobro.

160. *Zapopan: Economía, Empleo, equidad, Calidad de Vida, Educación, salud y seguridad pública.* Data México. (n.d.). https://www.economia.gob.mx/datamexico/es/profile/geo/zapopan.
161. Unfccc.int. Accessed July 6, 2024. https://unfccc.int/climate-action/un-global-climate-action-awards/climate-leaders/pacmetro.
162. *Climate-data.org*. Search results. (n.d.-a). https://en.climate-data.org/search/?q=Zapopan.
163. Bozzo Rotondo, L. "La Torre Cube en Zapopan (México)." *Hormigón y Acero* 56, no. 237 (2005).
164. Minkel, C., Schneider, Ronald Milton and Leite, Aureliano. "São Paulo." Encyclopedia Britannica, February 22, 2023. https://www.britannica.com/place/Sao-Paulo-Brazil.
165. Castillo-García, Rodolfo Francisco. "Achieve sustainable urban development in Lima Callao Megalopolis, Peru by 2050." *Eco Cities* 2, no. 2 (2021): 15.
166. Castillo-García, Rodolfo Francisco. "Achieve sustainable urban development in Lima Callao Megalopolis, Peru by 2050." *Eco Cities* 2, no. 2 (2021): 15.
167. *Climate-data.org*. Melbourne climate: Weather Melbourne & temperature by month. (n.d.-a). https://en.climate-data.org/oceania/australia/victoria/melbourne-764100/.
168. Fracalossi, I. (2014, June 4). *João Moura Building/Nitsche arquitetos*. ArchDaily. https://www.archdaily.com/511979/joao-moura-building-nitsche-arquitetos.
169. "Barranco." Encyclopedia Britannica. Accessed July 27, 2023. https://www.britannica.com/place/Barranco-Peru.
170. "Barranco." Encyclopedia Britannica. Accessed July 27, 2023. https://www.britannica.com/place/Barranco-Peru.
171. "Urban Agriculture in Peru Is Creating Communities Committed to Healthy Living and a Cleaner Environment." Cuso International, May 11, 2023. https://cusointernational.org/stories/urban-agriculture-in-peru-is-creating-communities-committed-to-healthy-living-and-a-cleaner-environment/.
172. *Climate-data.org*. Search results. (n.d.-a). https://en.climate-data.org/search/?q=lima.
173. Universita Luigi Bocconi, Milan - grafton architects. Accessed July 5, 2024. https://www.graftonarchitects.ie/universita-luigi-bocconi-milan.
174. Rio de Janeiro population 2024. Accessed July 6, 2024. https://worldpopulationreview.com/world-cities/rio-de-janeiro-population.
175. Colin, Benoit. "4 Inspirations for Sustainable Transport from Rio de Janeiro." World Resources Institute, March 23, 2015. https://www.wri.org/insights/4-inspirations-sustainable-transport-rio-de-janeiro.
176. *Climate-data.org*. Search results. (n.d.-a). https://en.climate-data.org/search/?q=Rio%2Bde%2BJaneiro.
177. RB12 - Triptyque Architecture. Accessed July 6, 2024. https://triptyque.com/en/project/rb12-2/.
178. Medellin population 2024. Accessed July 6, 2024. https://worldpopulationreview.com/world-cities/medellin-population.
179. Galvin, Madeleine, and Anne Maassen. "Urban Transformations: In Medellín, Metrocable Connects People in More Ways than One." World Resources Institute, March 20, 2019. https://www.wri.org/insights/urban-transformations-medellin-metrocable-connects-people-more-ways-one.
180. Galvin, Madeleine, and Anne Maassen. "Urban Transformations: In Medellín, Metrocable Connects People in More Ways than One." World Resources Institute, March 20, 2019. https://www.wri.org/insights/urban-transformations-medellin-metrocable-connects-people-more-ways-one.
181. *Climate-data.org*. Medellín climate: Weather Medellín & temperature by month. (n.d.). https://en.climate-data.org/south-america/colombia/antioquia/medellin-4088/.
182. Empresa de Desarrollo Urbano EDU + Salmaan Craig. "How to Design a 'Building That Breathes': A Sustainable Case Study of Colombia's Edu Headquarters." ArchDaily, June 9, 2017. https://www.archdaily.com/872018/how-to-design-a-building-that-breathes-a-sustainable-case-study-of-colombias-edu-headquarters.

183. “Edu Headquarters: A New Kind of Ventilation, a New Kind of Experiment.” urbanNext. Accessed July 6, 2024. https://urbannext.net/edu-headquarters-ventilation-experiment/.
184. “Hippodrome Tower: OAB - Office of Architecture in Barcelona.” Archello. Accessed July 6, 2024. https://archello.com/project/hippodrome-tower.
185. “Hippodrome Tower: OAB - Office of Architecture in Barcelona.” Archello. Accessed July 6, 2024. https://archello.com/project/hippodrome-tower.
186. Colombiareports. (2023, January 28). *Bogota: Colombia reports*. Colombia News | Colombia Reports. https://colombiareports.com/bogota/.
187. Laura Roteis managing editor of gb&d magazine. “Bogotá Is on a Mission to Become a More Sustainable City - GB&D.” gb&d magazine, September 4, 2020. https://gbdmagazine.com/bogota/.
188. *Climate-data.org*. Search results. (n.d.-a). https://en.climate-data.org/search/?q=Bogot%C3%A1.
189. “Universidad EAN.” William McDonough + Partners, May 18, 2023. https://mcdonoughpartners.com/projects/universidad-ean/.
190. “Atrio WSP Colombia.” WSPglobal. Accessed July 27, 2023. https://www.wsp.com/es-pe/proyectos/ed-co-atrio-wspcolombia.
191. Rshp. “Atrio – Office – Projects.” RSHP. Accessed July 5, 2024. https://rshp.com/projects/office/atrio/.
192. “Atrio WSP Colombia.” WSPglobal. Accessed July 27, 2023. https://www.wsp.com/es-pe/proyectos/ed-co-atrio-wspcolombia.
193. Harare population 2024. Accessed July 6, 2024. https://worldpopulationreview.com/world-cities/harare-population.
194. Harare, Z. A. C. T. “Zimbabwe.” *INDACOCHEA, FERNANDO J, Miami, FL.* 361 (1998): 114.
195. Harare, Z. A. C. T. “Zimbabwe.” *INDACOCHEA, FERNANDO J, Miami, FL.* 361 (1998): 114.
196. *Climate-data.org*. Harare climate: Weather Harare & temperature by month. (n.d.). https://en.climate-data.org/africa/zimbabwe/harare-province/harare-3530/.
197. Amr, Hosam M. Abd El Aziz, Nouran M. Ibrahim, and Hesham Sameh. “Biomimicry as a tool for minimizing energy consumption and improvement of thermal comfort: The case of office buildings.” In *International Conference for Sustainable Design of the Built Environment-SDBE London*, p. 376. 2018.
198. Accra population 2024. Accessed July 6, 2024. https://worldpopulationreview.com/world-cities/accra-population.
199. Arn, Jack. “Third world urbanization and the creation of a relative surplus population: A history of Accra, Ghana to 1980.” *Review (Fernand Braudel Center)* (1996): 413–443.
200. “Accra, Ghana: Greening and Beautification Project.” UrbanShift, September 25, 2023. https://www.shiftcities.org/publication/accra-ghana-greening-and-beautification-project.
201. *Climate-data.org*. Accra climate: Average Temperature by month, Accra water temperature. (n.d.). https://en.climate-data.org/africa/ghana/greater-accra-region/accra-534/.
202. “One Airport Square.” Mario Cucinella Architects. Accessed July 6, 2024. https://www.mcarchitects.it/en/projects/one-airport-square.
203. Addis Ababa population 2024. Accessed July 6, 2024. https://worldpopulationreview.com/world-cities/addis-ababa-population.
204. Ethiopia: Addis Ababa Urban Profile - UN-Habitat. Accessed July 5, 2024. https://unhabitat.org/sites/default/files/download-manager-files/Ethiopia Addis Ababa Urban Profile.pdf.
205. “Climate-Data.Org.” Climate Addis Ababa: Temperature, climate graph, Climate table for Addis Ababa. Accessed July 5, 2024. https://en.climate-data.org/africa/ethiopia/addis-ababa-1503/.
206. Rojas, Cristobal. “Lideta Market/Vilalta Studio.” ArchDaily, June 15, 2016. https://www.archdaily.com/789535/lideta-market-vilalta-arquitectura.
207. JOHN, GKP. “Mathematics of architecture: A study of the philosophy of emerging architecture.” (2016).

208. Lagos population 2024. Accessed July 6, 2024. https://worldpopulationreview.com/world-cities/lagos-population.
209. Agbola, Tunde, and Elijah M. Agunbiade. "Urbanization, slum development and security of tenure: The challenges of meeting millennium development goal 7 in Metropolitan Lagos, Nigeria." *Urban population-environment dynamics in the developing world: Case studies and lessons learned* (2009): 77–106.
210. "Climate-Data.Org." Lagos climate: Average Temperature by month, Lagos water temperature. Accessed July 5, 2024. https://en.climate-data.org/africa/nigeria/lagos/lagos-552/.
211. "Kingsway Tower." SAOTA Architecture and Design, April 16, 2022. https://www.saota.com/project/kingsway-tower/#2.
212. Casablanca population 2024. Accessed July 6, 2024. https://worldpopulationreview.com/world-cities/casablanca-population.
213. Cohen, Jean-Louis. "Architectural history and the colonial question: Casablanca, Algiers and beyond." *Architectural History* 49 (2006): 349–368.
214. "Climate-Data.Org." Casablanca climate: Average Temperature by month, Casablanca water temperature. Accessed July 5, 2024. https://en.climate-data.org/africa/morocco/casablanca/casablanca-164/.
215. Morphosis. Accessed July 6, 2024. https://www.morphosis.com/architecture/226/.
216. Johannesburg population 2024. Accessed July 6, 2024. https://worldpopulationreview.com/world-cities/johannesburg-population.
217. Beavon, Keith. *Johannesburg: The making and shaping of the city*. Vol. 9. Brill, 2022.
218. "Climate-Data.Org." Johannesburg climate: Weather Johannesburg & temperature by month. Accessed July 5, 2024. https://en.climate-data.org/africa/south-africa/gauteng/johannesburg-3221/.
219. "The Bank, Rosebank, Johannesburg." DAFFONCHIO & ASSOCIATES ARCHITECTS. Accessed July 6, 2024. https://www.daffonchio.co.za/the-bank-rosebank-johannesburg.
220. Melbourne population 2024. Accessed July 6, 2024. https://worldpopulationreview.com/world-cities/melbourne-population.
221. Davison, Graeme. "Australian urban history: a progress report." *Urban History* 6 (1979): 100–109.
222. "Climate-Data.Org." Melbourne climate: Weather Melbourne & temperature by month. Accessed July 5, 2024. https://en.climate-data.org/oceania/australia/victoria/melbourne-764100/.
223. Morris-Nunn, Robert. "CH2 [The City of Melbourne's new office building by Mick Pearce and DesignInc is Australia's first building based on biomimicry.]." *Architecture Australia* 96, no. 1 (2007): 90–99.
224. Bessel, C. "Violent Winds Pound Tokyo, Topple Mid-Construction Nine-Story Building." *SORA News*, April 18, 2016. https://soranews24.com/2016/04/18/violent-winds-pound-tokyo-and-topple-mid-construction-nine-story-building-%E3%80%90video%E3%80%91/.
225. "Tivoli Towers." *Skyscraper Center.* https://www.skyscrapercenter.com/building/tivoli-towers/18628.
226. "CBS News. 'Façade Partially Collapses at Crown Heights Apartment Building.' November 4, 2022. Accessed January 20, 2023. https://www.cbsnews.com/newyork/news/tivoli-towers-crown-heights-facade-collapse/."
227. Sky News. "Turkey-Syria Earthquake: Building Facade Collapses in Diyarbakir, Turkey." February 6, 2023. Available: https://news.sky.com/video/turkey-syria-earthquake-building-facade-collapses-in-diyarbakir-turkey-12804622.
228. Lahbib, P. "Why You Should Consider Aluminium Facades to Help Withstand Earthquakes." *Shapes*, November 8, 2020. Accessed April 23, 2023. https://www.shapesbyhydro.com/en/design-thinking/why-you-should-consider-aluminium-facades-to-help-withstand-earthquakes/.
229. Loads, Minimum Design. "Associated criteria for buildings and other structures." In *American Society of Civil Engineers*, pp. 6–15. 2017.
230. American Society of Civil Engineers ASCE 7-2010, "Minimum design loads for buildings and other structures." ASCE 7-10, ASCE, Reston, VA.

231. ASTM International. *Annual book of ASTM standards.* ASTM international, 2024.
232. (ICC), I. C. C. (n.d.). *2021 International Building Code (IBC): ICC Digital Codes.* 2021 INTERNATIONAL BUILDING CODE (IBC) | ICC DIGITAL CODES.
233. (ICC), I. C. C. (n.d.-b). Digital codes. https://codes.iccsafe.org/s/IBC2018/chapter-16-structural-design/IBC2018-Ch16-Sec1604.3#:~:text=URL%3A%20https%3A%2F%2Fcodes.iccsafe.org%2Fs%2FIBC2018%2Fchapter,100.
234. Besjak, Charles, Preetam Biswas, Georgi I. Petrov, Matthew Streeter, and Austin Devin. "Effects of Perimeter to Core Connectivity on Tall Building Behavior." *International journal of high-rise buildings* 6, no. 1 (2017): 1–9.
235. Grenfell Tower Inquiry, *Grenfell Tower Inquiry: Phase 1 Report* (London: UK Government, 2019), https://www.grenfelltowerinquiry.org.uk/phase-1-report.
236. Buchanan, A. H., and D. Barber. "Fire resistance of epoxied steel rods in glulam timber." In *Proceedings of Pacific timber engineering conference, Gold Coast, Australia.* 1994.
237. (ICC), International Code Council. "2021 International Building Code (IBC): ICC Digital Codes." 2021 INTERNATIONAL BUILDING CODE (IBC) | ICC DIGITAL CODES. Accessed June 2, 2024. https://codes.iccsafe.org/content/IBC2021P2/chapter-6-types-of-construction.
238. Brannigan, Francis L. Building construction *for the fire service.* Delmar Thomson Learning, 1992.
239. (ICC), International Code Council. "2021 International Building Code (IBC): ICC Digital Codes." 2021 INTERNATIONAL BUILDING CODE (IBC) | ICC DIGITAL CODES. Accessed June 2, 2024. https://codes.iccsafe.org/content/IBC2021P2/chapter-7-fire-and-smoke-protection-features#IBC2021P2_Ch07_Sec703.
240. ASTM, E. "ASTM E119-16a. Standard Test Methods for Fire Tests of Building Construction and Materials." ASTM International 21 (2016).
241. ASTM, E. "Standard test method for surface burning characteristics of building materials." (84).
242. McKenna, Sean T., Nicola Jones, Gabrielle Peck, Kathryn Dickens, Weronika Pawelec, Stefano Oradei, Stephen Harris, Anna A. Stec, and T. Richard Hull. "Fire behaviour of modern façade materials–Understanding the Grenfell Tower fire." *Journal of hazardous materials* 368 (2019): 115–123.
243. Stec, Anna A., and T. Richard Hull. "Assessment of the fire toxicity of building insulation materials." *Energy and Buildings* 43, no. 2–3 (2011): 498–506.
244. Delichatsios, M. A. "Enclosure and Façade fires: Physics and applications." *Fire Safety Science* 11 (2014): 3–27.
245. Henshell, Justin. *The manual of below-grade waterproofing*. Routledge, 2016. p.1–13.
246. ASTM. (n.d.). *Detailed overview - overview - about Us.* Detailed overview - Overview - About Us. https://www.astm.org/about/overview/detailed-overview.html.
247. Henshell, Justin. *The manual of below-grade waterproofing*. Routledge, 2016. p.12–3.
248. ASTM. "ASTM E 2556 Standard Specification for Vapor Permeable Flexible Sheet Water-Resistive Barriers Intended for Mechanical Attachment." ASTM International (2022).
249. Hill, John. *Guide to Chicago's Twenty-first-century Architecture.* University of Illinois Press, 2021.
250. Slanger, Dan. "Empire State Building Retrofit Surpasses Energy Savings Expectations." RMI, March 2, 2022. https://rmi.org/blog_empire_state_retrofit_surpasses_energy_savings_expectations/.
251. Handbook, ASHRAE Fundamentals. "ASHRAE Fundamentals Handbook." (1997).
252. Handbook, ASHRAE Fundamentals. "ASHRAE Fundamentals Handbook." (1997).
253. (ICC), International Code Council. "2021 International Energy Conservation Code (IECC): ICC Digital Codes." 2021 INTERNATIONAL ENERGY CONSERVATION CODE (IECC) | ICC DIGITAL CODES. Accessed June 2, 2024. https://codes.iccsafe.org/content/IECC2021P2/chapter-4-ce-commercial-energy-efficiency.
254. (ICC), International Code Council. "2021 International Energy Conservation Code (IECC): ICC Digital Codes." 2021 INTERNATIONAL ENERGY CONSERVATION CODE (IECC)

| ICC DIGITAL CODES. Accessed June 2, 2024. https://codes.iccsafe.org/content/IECC2021P2/chapter-4-ce-commercial-energy-efficiency.
255. Handbook, ASHRAE Fundamentals. "ASHRAE Fundamentals Handbook." (2005). P.25–1.
256. American Society for Testing and Materials. Committee D16 on Thermal Insulation. C1363-24 *Standard test method for thermal performance of building materials and envelope assemblies by means of a hot box apparatus.* ASTM International, 2024.
257. ASTM International. *ASTM E903-20, Standard Test Method for Solar Absorptance, Reflectance, and Transmittance of Materials Using Integrating Spheres.* ASTM International, 2020.
258. ASTM International. ASTM E 2178-21 Standard Test Method for Determining Air Leakage Rate and Calculation of Air Permeance of Building Materials. ASTM International, 2021.
259. S. A. Al-Sanea, M. F. Zedan, A. M. Al-Mujahid and Z. A. Al-Suhaibani, "Optimum R-values of building walls under different climatic conditions in the Kingdom of Saudi Arabia," *Applied Thermal Engineering,* vol. 96, pp. 92–106, 2016.
260. M. Bojic, F. Yik, K. Wan and J. Burnett, "In uence of envelope and partition characteristics on the space cooling of high-rise residential buildings in Hong Kong," *Building and Environment,* vol. 37, p. 347–355, 2002.
261. M. Bojic´ and F. Yik, "Cooling energy evaluation for high-rise residential buildings in Hong Kong," *Energy and Buildings,* vol. 37, p. 345–351, 2005.
262. W. A. Friess, K. Rakhshan, T. A. Hendawi and S. Tajerzadeh, "Wall insulation measures for residential villas in Dubai: A case study in energy efficiency," *Energy and Buildings,* vol. 44, pp. 26–32, 2012.
263. H. Ge and F. Baba, "Dynamic effect of thermal bridges on the energy performance of a low-rise residential building," *Energy and Buildings,* vol. 105, pp. 106–118, 2015.
264. M. Altaf and F. Hill, "Identifying Optimum Glazing Properties for Conserving Energy in Hot Semi-Arid Climate Regions," *Journal of Green Building,* vol. 16, no. 1, pp. 91–101, 2021.
265. Y. J. Lee, S. H. Kim, J. H. Ryu and K. H. Lee, "OptimizingWindow Glass Design for Energy Efficiency in South Korean Office Buildings: A Hierarchical Analysis Using Energy Simulation," *Buildings,* vol. 13, no. 11, 2023.
266. N. L. Sbar, L. Podbelski, H. M. Yang and B. Pease, "Electrochromic dynamic windows for office buildings," *International Journal of Sustainable Built Environment,* vol. 1, pp. 125–139, 2012.
267. Y. Ko, H. Oh, H. Hong and J. Min, "Energy Consumption Verification of SPD Smart Window, Controllable According to Solar Radiation in South Korea," *Energies,* vol. 13, no. 21, 2020.
268. Uno, Tomoko, Shuichi Hokoi, Sri Nastiti N. Ekasiwi, and Noor Hanita Abdul Majid. "Reduction of energy consumption by AC due to air tightness and ventilation strategy in residences in hot and humid climates." *Journal of Asian Architecture and Building Engineering* 11, no. 2 (2012): 407–414.
269. Liu, Yuan, Baizhan Li, and Xinyun Cao. "The research on the influence of building air tightness to energy consumption of residential building in a hot summer and cold winter zone in China." In *IOP Conference Series: Materials Science and Engineering*, vol. 609, no. 7, p. 072003. IOP Publishing, 2019.
270. Bjarke Ingels Group (BIG). "Shenzhen Energy Mansion HQ." *BIG.* Accessed July 8, 2024. https://big.dk/projects/shenzhen-energy-mansion-hq-2314.
271. Güneralp, Burak, and Karen C. Seto. "Environmental impacts of urban growth from an integrated dynamic perspective: A case study of Shenzhen, South China." *Global Environmental Change* 18, no. 4 (2008): 720–735.
272. Li, Yan, Hongwu Du, and Ceren Sezer. "Sky Gardens, Public Spaces and Urban Sustainability in Dense Cities: Shenzhen, Hong Kong and Singapore." *Sustainability* 14, no. 16 (2022): 9824. https://doi.org/10.3390/su14169824.
273. Shenzhen population 2024. Accessed July 8, 2024. https://worldpopulationreview.com/world-cities/shenzhen-population.
274. Hu, Richard. "The state of smart cities in China: The case of Shenzhen." *Energies* 12, no. 22 (2019): 4375.

275. Luo, Keyu, Zhenyu Wang, Wei Sha, Jiansheng Wu, Hongliang Wang, and Qingliang Zhu. "Integrating sponge city concept and neural network into land suitability assessment: evidence from a satellite town of shenzhen metropolitan area." *Land* 10, no. 8 (2021): 872.
276. "Climate-Data.Org." Shenzhen climate: Average Temperature by month, Shenzhen water temperature. Accessed July 8, 2024. https://en.climate-data.org/asia/china/guangdong/shenzhen-1064927/.
277. Büro Ole Scheeren. "Mahanakhon." Büro Ole Scheeren. Accessed July 8, 2024. https://buro-os.com/projects/mahanakhon.
278. Ilgin, Hüseyin Emre. "A Search for a New Tall Building Typology: Structural Hybrids." *Proceedings of the LivenARCH* (2021).
279. Summ, Wong Mun, and Richard Hassell. "Sustainable Building Project Report: Tall Buildings in Southeast Asia–A Humanist Approach to Tropical Highrise." *International Journal of Sustainable Building Technology and Urban Development* 2, no. 1 (2011): 21–28.
280. Oldfield, Philip. *The Sustainable Tall Building: A Design Primer*. Routledge, 2019.
281. Bangkok population 2024. Accessed July 8, 2024. https://worldpopulationreview.com/world-cities/bangkok-population.
282. Mateo-Babiano, Iderlina B. "Public life in Bangkok's urban spaces." *Habitat International* 36, no. 4 (2012): 452–461.
283. Vichiensan, Varameth, Nattapon Suk-kaew, Mass Rapid Transit Authority, Masanobu Kii, Yoshitsugu Hayashi, and Global Smart City. "TOD Factors Influencing Urban Railway Ridership in Bangkok."
284. "Climate-Data.Org." Bangkok climate: Weather Bangkok & temperature by month. Accessed July 8, 2024. https://en.climate-data.org/asia/thailand/bangkok/bangkok-6313/.
285. Pilea. Le Nouvel, Kuala Lumpur, December 1, 2023. https://www.verticalgardenpatrickblanc.com/realisations/kuala-lumpur/le-nouvel-kuala-lumpur.
286. SSYENC. "Completed Project Portfolio." Accessed July 8, 2024. http://www.ssyenc.co.kr/file/story/html/story3/V3(영)6Completed Project 1.pdf.
287. Kuala Lumpur population 2024. Accessed July 7, 2024. https://worldpopulationreview.com/world-cities/kuala-lumpur-population.
288. "Climate-Data.Org." Kuala Lumpur climate: Weather Kuala Lumpur & temperature by month. Accessed July 7, 2024. https://en.climate-data.org/asia/malaysia/kuala-lumpur/kuala-lumpur-715107/.
289. "Climate-Data.Org." Kuala Lumpur climate: Weather Kuala Lumpur & temperature by month. Accessed July 7, 2024. https://en.climate-data.org/asia/malaysia/kuala-lumpur/kuala-lumpur-715107/.
290. Yusoff, Nurulhuda Abdul Hamid, Anuar Mohd Noor, and Rosmadi Ghazali. "City skyline conservation: sustaining the premier image of Kuala Lumpur." *Procedia Environmental Sciences* 20 (2014): 583–592.
291. Halawa, Edward, Amirhosein Ghaffarianhoseini, Ali Ghaffarianhoseini, Jeremy Trombley, Norhaslina Hassan, Mirza Baig, Safiah Yusmah Yusoff, and Muhammad Azzam Ismail. "A Review on Energy Conscious Designs of Building Façades in Hot and Humid Climates: Lessons for (and from) Kuala Lumpur and Darwin." *Renewable and* Sustainable Energy Reviews 82 (2018): 2147–2161.
292. Al-Kodmany, K. "Greenery-Covered Tall Buildings: A Review." *Buildings* 13, no. 9 (2023): 2362. https://doi.org/10.3390/buildings13092362.